AF167510

SpringerBriefs in Ethics

More information about this series at http://www.springer.com/series/10184

Seumas Miller

Dual Use Science and Technology, Ethics and Weapons of Mass Destruction

 Springer

Seumas Miller
Charles Sturt University
Canberra
Australia

and

TU Delft
The Hague
Netherlands

and

University of Oxford
Oxford
UK

ISSN 2211-8101   ISSN 2211-811X   (electronic)
SpringerBriefs in Ethics
ISBN 978-3-319-92605-6   ISBN 978-3-319-92606-3   (eBook)
https://doi.org/10.1007/978-3-319-92606-3

Library of Congress Control Number: 2018942922

Printed on acid-free paper

This Springer imprint is published by the registered company Springer International Publishing AG
part of Springer Nature
The registered company address is: Gewerbestrasse 11, 6330 Cham, Switzerland

Acknowledgements

Seumas Miller (Australian Graduate School of Policing and Security at Charles Sturt University, the Department of Values, Technology and Innovation at Delft University of Technology (TU Delft) and the Uehiro Centre for Practical Ethics at the University of Oxford) and Jonas Feltes (ERC Advanced Grant on Global Terrorism and Collective Moral Responsibility: Redesigning Military, Police and Intelligence Institutions in Liberal Democracies, Delft University of Technology (TU Delft)) co-authored Chap. 5; Seumas Miller and Behnam Taebi (Department of Values, Technology and Innovation, Delft University of Technology (TU Delft)) co-authored Chap. 6; Seumas Miller and Terry Bossomaier (Charles Sturt University) co-authored Chap. 7. This project was partly funded by the ERC Advanced Grant on Global Terrorism and Collective Moral Responsibility: Redesigning Military, Police and Intelligence Institutions in Liberal Democracies (www.counterterrorismethics.com) held by Seumas Miller at the Department of Values, Technology and Innovation at Delft University of Technology (TU Delft) and the Uehiro Centre for Practical Ethics at the University of Oxford.

Thanks to the editors of the following academic publications for the use of some of the material of Seumas Miller contained therein: "Academic Autonomy", in C. A. J. Coady (ed.) *Why Universities Matter*, Sydney: Allen and Unwin, 2000; "Ethical and Philosophical Consideration of the Dual Use Dilemma in the Biological Sciences" (with Michael Selgelid), *Science and Engineering Ethics* vol. 13 2007; *Ethical and Philosophical Consideration of the Dual Use Dilemma in the Biological Sciences* (with Michael Selgelid), Springer, 2008; *Report on Biosecurity and Dual Use Research* (with Koos van der Bruggen and Michael Selgelid), The Hague: Dutch Research Council, 2011; "Moral Responsibility, Collective Action Problems and the Dual Use Dilemma in Science and Technology" in Brian Rappert and Michael Selgelid (eds.), *On the Dual Uses of Science and Ethics: Principles, Practices and Prospects*, Canberra: ANU Press, 2013; "Ignorance, Technology and Collective Responsibility" in Rik Peels (ed.), *Perspectives on Ignorance from Moral and Social Philosophy,* Oxford: Routledge, 2017.

Contents

Chapter 1
Introduction

Abstract The problem of dual-use science research and technology arises because such research and technology has the potential to be used for great evil as well as for great good. On the one hand, knowledge is a necessary condition, and perhaps a constitutive feature, of technologies that contribute greatly to individual and collective well-being. Consider, for example, nuclear technology that enables the generation of low cost electricity in populations without obvious alternative energy sources. So technological knowledge is a good thing and ignorance of it a bad thing. On the other hand, these same technologies can be extremely harmful to individuals and collectives. Consider, for example, the atomic bombs dropped on Hiroshima and Nagasaki. So it seems that, at least with respect to some technologies, knowledge is a bad thing and ignorance a good thing. Accordingly, the question arises as to whether we ought to limit scientific research and/or the development of technology and, if so, which research or technology, in what manner and to what extent.

The problem of dual-use science research and technology arises because such research and technology has the potential to be used for great evil as well as for great good.[1] On the one hand, knowledge is a necessary condition, and perhaps a constitutive feature, of technologies that contribute greatly to individual and collective well-being. Consider, for example, nuclear technology that enables the generation of low cost electricity in populations without obvious alternative energy sources. So technological knowledge is a good thing and ignorance of it a bad thing. On the other hand, these same technologies can be extremely harmful to individuals and collectives. Consider, for example, the atomic bombs dropped on Hiroshima and Nagasaki. So it seems that, at least with respect to some technologies, knowledge is a bad thing and ignorance a good thing. Accordingly, the question arises as to whether we ought to limit scientific research and/or the development of technology and, if so, which research or technology, in what manner and to what extent.

Evidently scientific knowledge that enables the development of dual use technologies is potentially dangerous and therefore, where possible, it should be restricted

---

[1] See, for example, Miller and Selgelid (2007, pp. 523–580), Rappert and Selgelid (2013), Meier and Hunger (2014) and Tucker (2012).

© The Author(s) 2018   1
S. Miller, *Dual Use Science and Technology, Ethics and Weapons of Mass
Destruction*, SpringerBriefs in Ethics, https://doi.org/10.1007/978-3-319-92606-3_1

or perhaps even not acquired in the first place. In short, contrary to popular opinion, there ought to be a degree of *collective scientific ignorance*, at least among members of the general population. But what is collective ignorance and how does it relate to collective knowledge? More generally, dual use science research and technology are collective epistemic or knowledge-aiming enterprises that produce collective benefits but can also at times cause collective harms. Indeed, they are enterprises conducted by institutions, such as universities, private sector firms and military organisations. Naturally, if the benefits are to flow there is a need to protect and promote scientific freedom. On the other hand, in relation to the potential for harm, scientists and others have a moral responsibility, even if not a legal responsibility, to cooperate in order to avert or, at least, minimise the risks; so dual use research and technology is a matter of *collective moral responsibility*. But what is collective responsibility and how does it figure in the varied scientific and institutional contexts of the collective epistemic enterprises in question? More specifically, should some dual use research be impermissible or, if not, should access to the resulting scientific knowledge be highly restricted, e.g. censored? What institutional arrangements, e.g. regulations, ought to be put in place in relation to dual use research? These are the questions that this work seeks to address.

Chapters 2, 3 and 4 are theoretical in character and could be skipped by those uninterested in the theoretical aspects of the problem. In Chap. 2 the key concept of dual use is defined. In Chap. 3 analyses of collective knowledge and collective ignorance are proffered. In Chap. 4 a theory of collective responsibility is presented. Chapters 5, 6, 7 and 8 each focus on a particular scientific field or industry of dual use concern, namely, the chemical industry, the nuclear industry, cyber-technology and the biological sciences (respectively).

The problem of dual-use research and technology arises in its most obvious form in the context of weapons of mass destruction (WMDs), whether chemical, nuclear, cyber or biological weapons. Scientific research originally conducted for beneficial peaceful purposes has also enabled WMDs. Moreover, the problem has been exacerbated by the growth of international terrorist groups, such as Al Qaeda and ISIS (Islamic State of Iraq and Syria), who evidently would be willing to use WMDs, if they could get their hands on them. Indeed, ISIS has already used chemical weapons in Iraq (as has their protagonist in Syria, the Assad regime). The use of chemical weapons in World War 1 and atomic weapons in World War 2 graphically illustrated the problem of dual use science and technology. In the biological sciences the dual use problem has arisen in its most acute form in relation to recent advances in synthetic biology which have enabled the creation of pathogens de novo. Unfortunately, this important scientific breakthrough has a downside; the potential for a 'superbug' pandemic. More specifically, this recent research includes gain of function (GOF) research, e.g. research that enables highly virulent pathogens to possess increased transmissibility to humans. Another area of dual use concern is new and emerging cyber-technology, including the development and deployment of computer viruses to engage in denial of service attacks that may well put lives at risk by, for instance, disabling life support systems in hospitals.

References

Meier, Oliver, and Iris Hunger. 2014. *Between Control and Cooperation: Dual-use, Technology Transfers and the Non-Proliferation of Weapons of Mass Destruction*. Osnabruck: DSF.

Miller, Seumas, and Michael Selgelid. 2007. Ethical and Philosophical Consideration of the Dual Use Dilemma in the Biological Sciences. *Science and Engineering Ethics* 13: 523–580.

Rappert, Brian, and Michael Selgelid (eds.). 2013. *On the Dual Uses of Science and Ethics: Principles, Practices and Prospects*. Canberra: ANU Press.

Tucker, J.B. (ed.). 2012. *Innovation, Dual Use, and Security: Managing the Risks of Emerging Biological and Chemical Technologies*. Harvard: MIT Press.

Chapter 2
Concept of Dual Use

Abstract There are a number of different preliminary definitions of dual use familiar in the literature. Research or technology is dual use if it can be used for both: (1) Military and civilian (i.e. non-military) purposes; (2) Beneficial and harmful purposes—where the harmful purposes are to be realised by means of Weapons of Mass Destruction (WMDs); (3) Beneficial and harmful purposes—where either the harmful purposes involve the use of weapons as means, and usually WMDs in particular, or the harm aimed at is on a large-scale but does not necessarily involve weapons or weaponisation. I favour the third definition of "dual use"—at least as a preliminary definition—since some dual use research, such as Gain of Function research in the biological sciences, need not involve a process of weaponisation or a military purpose. However, further conceptual unpacking is called for and provided in this chapter.

2.1 Definition of Dual Use Science and Technology

As noted in Chap. 1, the expression "dual use" refers to scientific research or technology that can be used for both beneficial/good and harmful/bad purposes.[1] However, this general sense of dual use is too broad since it has the effect that almost everything could count as dual use. For instance, machetes are used for farming, but they were also used in the Rwandan genocide in 1994 as tools of murder. So we require a narrower notion of dual use. Most of the current debate has focused on research and technologies with implications not simply for weapons but for weapons of mass destruction (WMDs), in particular—i.e., where the harmful consequences of malevolent use would be on an extremely large scale (and, likewise, the benefits of benevolent use would be large-scale). That said, as mentioned in the Chap. 1,

---

[1] See, for example: Miller and Selgelid (2007), van der Bruggen et al. (2011, 1–122), Miller (2013), Meier and Hunger (2014) and Tucker (2012). Some material (as opposed to technologies), e.g. toxins, might be dual use if, for instance, they are not naturally occurring but were man-made. However, for the sake of simplicity I will not refer to dual use materials unless this is required in the particular case under discussion.

© The Author(s) 2018
S. Miller, *Dual Use Science and Technology, Ethics and Weapons of Mass Destruction*, SpringerBriefs in Ethics, https://doi.org/10.1007/978-3-319-92606-3_2

defining dual use simply in terms of WMDs yields too narrow a notion given, for instance, GOF research in the biological sciences[2] (see Chap. 8). Accordingly, let us try to get a better fix on a serviceable notion of dual use by setting out a number of different preliminary definitions of dual use familiar in the literature[3] and doing so on the assumption that any definition will involve a degree of stipulation.

Research or technology is dual use if it can be used for both:

1. Military and civilian (i.e. non-military) purposes;
2. Beneficial and harmful purposes—where the harmful purposes are to be realised by means of WMDs;
3. Beneficial and harmful purposes—where either the harmful purposes involve the use of weapons as means, and usually WMDs in particular, or the harm aimed at is on a large-scale but does not necessarily involve weapons or weaponisation.[4]

I favour the third definition of "dual use"—at least as a preliminary definition—since some dual use research, such as GOF research in the biological sciences, need not involve a process of weaponisation or a military purpose. However, further conceptual unpacking is called for.

(1) In relation to the *purposes* (or ends) of the research, we need to distinguish the following conceptual axes: (i) beneficial/harmful; (ii) military/non-military; and (iii) within the category of military purposes, the sub-categories of offensive/protective. Consider the aerosolisation of a pathogen undertaken for a military purpose. The purpose in question might be offensive, e.g. biowarfare; but it might simply be protective, e.g. to understand the nature and dangers of such aerosolisation in order to prepare protections against an enemy known to be planning to deploy the aerosolised pathogen in question as a weapon.

The categories beneficial/harmful and military/non-military do not necessarily mirror one another. Some non-military purposes are, nevertheless, harmful, e.g. the supplier of a vaccine releasing a pathogen to make large numbers of people sick in order that the sick buy the vaccine against the pathogen and, thereby, increase the supplier's profits. And some military purposes might be good, e.g. the above-mentioned research on the aerosolisation of a pathogen undertaken for purely protective purposes in the context of a morally justified war. The United States Project BioShield is an example of research aimed at providing "new tools to improve medical countermeasures protecting Americans against a chemical, biological, radiological or nuclear (CBRN) attack."[5] However, some of the protective research would probably yield results that could assist in the development and delivery of biological weapons.

---

[2]National Science Advisory Board for Biosecurity Framework for Conducting Risk and Benefit Assessments of Gain-of-Function Research (2015), Selgelid (2016).

[3]Rappert and Selgelid (2013).

[4]These definitions assume that the benefits are also on a large-scale. Moreover, there is a distinction between an object which is a weapon merely because used as one, e.g. a brick used to hit someone on the head, and a weapon which was designed as such from material which is not in itself useable as a weapon and, therefore, needs to go through a process of weaponisation, e.g. a biological agent used in a bioweapon.

[5]US Department of Health and Human Services (2004).

(2) Dual-use refers to two conceptually distinct groups[6]: (i) those who initially undertake the research or develop the technology (let us refer to these as original researchers); and (ii) those who use the results of the work of these original researchers for some purpose other than that intended by the original researchers (let us refer to these as secondary users). (Those who use the research for the purpose for which it was originally intended can now be referred to as primary users.) For example, the above-mentioned research on the aerosolisation of a pathogen (conducted by the original researchers) might be used for offensive purposes by those fighting an unjust war (the secondary users).

In relation to the term, "use", we can distinguish: (i) actually or potentially used in accordance with the purpose for which it was designed (design-purpose); (ii) actually or potentially used for some purpose other than that for which it was specifically designed; (iii) actually or potentially used for a benevolent and, therefore, morally good purpose; (iv) actually or potentially used for a malevolent and, therefore, morally bad purpose.[7] Dual-use dilemmas typically involve: (A) original researchers undertaking scientific research or developing technology for a good purpose—the design-purpose is good; and (B) malevolent secondary (actual or potential) users—the research is to be used to cause great harm. This is consistent with their being some other group of original researchers who had a malevolent design-purpose. However, on my definition of dual use there needs to be a group of original researchers who have a good purpose. This good purpose is either a good design-purpose or a morally neutral design-purpose which is a means to some further good purpose that they have.

Accordingly, the original researchers might have as a design-purpose to demonstrate how to render a vaccine against a highly transmissible pathogen ineffective. This design-purpose can itself be in the service of a benevolent further purpose, e.g. the purpose of enhancing the effectiveness of the vaccine. Alternatively, the achievement of this design-purpose could be used for a malevolent further purpose, e.g. to render the vaccine ineffective as a means to realise the goal of spreading the disease caused by the pathogen in question. According to my above favoured definition, the research in question is only dual use if there are some group of researchers with the good purpose of enhancing the effectiveness of the vaccine. Admittedly, this is a stipulation that could be resisted. However, it is not an unmotivated stipulation. Most academic discussions of dual use science and technology implicitly assume that there are original researchers with a benevolent purpose (and that there is another group—either researchers or secondary users—with malevolent purposes). It follows that there is now a presumption in favour of accepting this assumption in one's definition of dual use.

---

[6]Two things can be conceptually distinct even if under some description they are the same thing. Thus being married is conceptually distinct from being a scientist. However, Jones can be a married scientist. Similarly, the original researcher could also be the secondary user, notwithstanding that original researcher and secondary user are distinct concepts.

[7]I am assuming that in the final analysis the dual use dilemma is a moral dilemma and, therefore, the harms and benefits in question are morally significant (either directly or indirectly).

(3) In relation to the *avoidable*[8] *outcomes* of the scientific research or technology, we can distinguish: (i) intended outcomes; (ii) unintended but foreseen outcomes; (iii) unforeseen (but foreseeable) outcomes; and (iv) unforeseeable outcomes. An example of an unintended outcome is the spread of radiotoxic material into the environment from a damaged nuclear reactor resulting from a tsunami, as happened in Fukushima, Japan in 2011. However, such accidents are not *obviously* instances of the dual-use dilemma (although some might be—see below). For something to be an instance of a dual-use dilemma, both outcomes (the two horns of the dual-use dilemma) need to be (actually or potentially) intended (or at least foreseen or foreseeable) by someone; there needs to be two sets of (actual or potential) *users*. Naturally, an outcome might be unintended and unforeseen (even unforeseeable) by the original researcher or technologist but, nevertheless, intended by the secondary user. Thus, scientists who preserve a small number of smallpox samples for pure research purposes in the context of a policy of mandatory destruction of samples might not intend or foresee that they might be used for malevolent purposes by others, e.g. weaponised. Again, scientists who develop the process of nuclear fission to be used for power generation might not intend or foresee that the same process might be used to build atomic bombs.

Notice that in the case of GOF research, such as research that enables the creation of a highly virulent pathogen that is transmissible to humans, the researchers (presumably) do not have a malevolent purpose.[9] Perhaps they want to understand the process by means of which such a virulent pathogen might mutate and put humans at risk having as an ultimate end to create a vaccine against such a transmissible pathogen. Nevertheless, these researchers have in fact created a highly dangerous new pathogen which has the potential to be intentionally released into a human population by some secondary user. Accordingly, such GOF research is dual use on our definition, notwithstanding that it does not involve an explicit process of weaponisation.

Many, if not most, so-called dual use dilemmas are not really dilemmas in the narrow sense of being situations involving two options which are equally morally problematic. In the first place, the dilemmas in question could be tri-lemmas; indeed, there could be four or five or some very large number of options all of which are equally morally problematic. In the second place, the options are not generally *equally* morally problematic. Certainly, there are moral considerations for and against each of the options, however it may well be that, all things considered, one of the options is morally preferable to the others and that this is relatively obvious to any rational, morally sensitive person. The point is rather that there are at least some significant moral costs associated with each of the available options.

---

[8] I am assuming that the relevant outcomes of dual use research are avoidable even if only by refraining from conducting the research. I am further assuming that the scientists in question could have avoided conducting the research. This raises the question of scientists operating in authoritarian states who are coerced into conducting certain research but also of the possibility of individuals *jointly* avoiding some activity or outcome. See Chap. 4.

[9] Selgelid (2016).

As already noted many, if not most, scientific discoveries and, especially, new technologies, have dual use potential in the trivial sense that they could be used by someone for some malevolent purpose. Indeed, any newly designed object, such as the first baseball bat, has dual use potential in this trivial sense. After all, baseball bats can be used to hit people over the head, as well as for the enjoyment of playing baseball. However, it is implicit in the use of the term "dual use" in play in the academic literature that the potential harm in question is of a very great magnitude and able to be caused by a one-off action, e.g. the potential of atomic physics to lead to the creation of the hydrogen bomb, the potential of genetic engineering to lead to a super-virus. The contrast here is with the multiple acts of, say, numerous people killing numerous other people by hitting them on their heads with baseball bats, such that while thousands, let us implausibly assume, are killed in this protracted series of attacks a single act of hitting at best kills only one person.

Note that accidents involving science and technology, even accidents on a very large scale, such as the Union Carbide Bhopal chemical disaster and the Chernobyl and Fukushima nuclear disasters, are not *necessarily* dual use in our sense since there is no secondary evil user. More generally, questions of security should be conceptually demarcated from questions of safety. Nevertheless, such disasters might be dual use if they involve culpable negligence. Here two points need to be kept in mind. Firstly, if it is more or less predictable that there will be a *morally culpable large-scale harm-causing* secondary user of the science and technology in question then it may be dual use, notwithstanding that this secondary user did not *intend* to do evil. Perhaps there is gross negligence with respect to safety on the part of a secondary user (who might in fact also be the original researcher) leading to massive loss of life and this was foreseen (or, at least, reasonably foreseeable) by the original researchers. Accordingly, the line between safety and security is in practice blurred; it is blurred at the point at which there is culpable negligence. Culpable negligence is both a safety and a security issue; hence by my lights dual use issues while primarily matters of security are also to some extent matters of safety. Once again there is an element of stipulation here. However, I am seeking a concept of dual use that does not embrace unforeseeable accidents; surely an unforeseeable accident is not a *use* since it is not an *act* per se but rather an event. The notion of culpability serves my purpose here since, arguably, those who are culpably negligent have committed (in some sense) *acts* of omission. Secondly, the original research which enabled the construction of such industrial plants might be dual use. Thus the process of nuclear fission which has as a by-product highly radioactive fissile material may well be dual use, given the known risk of large-scale harm to humankind posed by such material.

2.2 Weapons of Mass Destruction (WMDs)

The history of science and technology is replete with examples of scientific research being used intentionally or unintentionally to create weapons, including WMDs. Scientists have developed chemical, nuclear, cyber and biological weapons. Such weapons include the following historical examples: the mustard gas used by German and British armies in World War 1; the aerial spraying of plague-infested fleas by the Japanese military in World War 2 that killed thousands of Chinese civilians; the dropping of atomic bombs on Hiroshima and Nagasaki by the US Air Force in World War 2; the large-scale biological weapons program in the Soviet Union from 1946 to 1992; the biological weapons program of the apartheid government in South Africa, and; the use of chemical agents by Saddam Hussein's Iraqi regime against Kurds in 1988 and by the Assad government against opposition forces in Syria in 2015. Recently, cyber-technology has been used to create cyber-weapons which have the potential to cause large-scale harm, e.g. malware used in denial of service attacks. Stuxnet was a worm used to disable Iran's nuclear facilities and, more recently, WannaCry disabled the UK National Health Service's computer systems.

In recent decades there have been a number of high profile 'defections' of scientists from developed liberal democratic states to authoritarian and/or less developed states with WMD programs. For example, Dr. Abdul Qadeer Khan joined, and in large part established, Pakistan's nuclear weapons program after working for Urenco in the Netherlands, and Frans van Anraat (also from the Netherlands) went to Iraq to assist Saddam Hussein's WMD program producing mustard gas.

In the light of the above, it would be naive to assume that the scientific community can be entirely trusted to regulate itself in relation to dual use problems. After all, thousands of scientists have worked in the above-mentioned and other WMD, e.g. chemical, nuclear, cyber and biological weapons, programs, and in doing so have had as their institutionally given collective ends the production of chemical, nuclear, cyber and biological weapons (respectively). Accordingly, these scientists are (presumptively, at least) directly collectively morally responsible for the existence of those weapons—even if most of these scientists individually only had a minor role and, therefore, only a small share of the overall collective responsibility should be attributed to each of these—and, in the case of scientists working for authoritarian governments, for enabling authoritarian regimes to possess them.[10] Moreover, on some occasions, as already noted, WMD's have actually been used; accordingly, the scientists involved in the development of these WMD's are morally implicated, even if only indirectly, in the harms caused by such use.

The security threat posed by WMDs involves various categories of harm and exponentially increases the magnitude of these harms. The 'harms' in question include not only physical and psychological harms to human beings, but also damage to material things, such as artefacts and the physical environment, damage to institutions and, for that matter, to computer software and the like. The security threat posed by WMDs is perhaps most obvious in the case of chemical and nuclear weapons. How-

---

[10]For an analysis of collective ends and collective responsibility see Chap. 4. As I argue in Chap. 4, it does not follow that these scientists are morally blameworthy since moral responsibility should be distinguished from blameworthiness, albeit the former presupposes the latter. Moreover, collective responsibility is not simply aggregate individual responsibility, so my reference to a share of collective responsibility here and elsewhere does not imply a simple numerical process of disaggregation of collective responsibility based on the numbers of participants in the joint action in question. As argued in Chap. 4, matters are more complex than that.

ever, as has become clear in recent times, the security threat posed by cyber-attacks involving computer viruses and the like is also very great, given the vulnerability to cyber-attacks of the ICT (information and communication technology) critical infrastructure relied upon by government departments, businesses, hospitals, police organisations and so on. Moreover, the security threat associated with infectious diseases relating to their potential use in biological weapons is a further case in point. The use of a highly contagious and deadly infectious disease in a biological weapon could lead to an epidemic with have catastrophic consequences.

The potential users of WMDs include not only state actors but also non-state actors, such as terrorist groups, nihilistic 'end-of-the-world' groups and, potentially, malevolent 'lone-wolf' actors. Of course the threat of the use of some WMDs by some kinds of malevolent actor is far greater than others. The military forces of nation-states with sophisticated R&D programs are far more likely to use nuclear weapons than non-state actors, at least in the near-term. On the other hand, the use of chemical weapons by non-state actors, such as international terrorist groups, is far more likely than is their use of nuclear weapons. This is in part because of the availability of stockpiles of the relevant toxins and in part because the delivery systems of chemical weapons are relatively unsophisticated.

Cyber-weapons provide a somewhat different kind of example since, as we shall see in Chap. 7, the harms that they cause are typically indirect and more diffuse than that of 'conventional' WMDs. Moreover, while state actors appear to be behind the more serious cyber-attacks thus far, it is far from obvious that non-state actors will not be perpetrators in the future, even if they have not been thus far.

Biological weapons are different again (see Chap. 8). Much debate regarding the threats posed by biological weapons—and bioterrorism in particular—has focused on the issue of dual-use life science research.[11] While advances in genetics, biotechnology, and synthetic biology may lead to important medical progress, they might also enable production of a new generation of biological weapons of mass destruction. Such dangers are well illustrated by recent research (conducted in the Netherlands and United States) that demonstrated how to produce a strain of avian influenza (H5N1) that is highly contagious among ferrets (which provide the best model for influenza among humans). Due to concerns about the public health and security implications of publishing details about this research, the US National Science Advisory Board for Biosecurity (NSABB), in December 2011, recommended that detailed description of materials and methods be omitted from publications (in science journals) describing the experiments in question.[12]

---

[11] See National Research Council (2004) and Miller and Selgelid (2007).

[12] The NSABB subsequently reversed its decision/recommendation in March 2012. See Selgelid (2016).

2.3   No Means to Harm (NMH) Principle

In the light of the above discussion we can identify the fundamental moral principle in play in dual use contexts, namely, the principle not to provide the means for harm to be done and, in particular, harm to human beings. Let us elaborate in general terms some of the implications of this principle for the so-called dual use dilemma.

As already indicated, in some cases it is, and ought to be, unlawful for scientists to provide others with the means to do great harm, e.g. scientists who develop chemical weapons. However, in the case of some dual use research it is neither feasible nor reasonable for that research or, at least, the dissemination of its findings to be unlawful; nevertheless, it might be morally desirable all things considered that the research in question not be undertaken or that its results not be disseminated. Especially in some of the dual use cases in which the research or its dissemination ought not to be unlawful, scientists and policymakers may face a moral dilemma.

Consider the above-mentioned dual use dilemma that arose in nuclear science in relation to the process of fission:

Option 1—Scientists morally ought to conduct research into nuclear fission since it enables the provision of a much needed source of power for civilian purposes.
Option 2—Scientists morally ought *not* to have undertaken the research into nuclear fission since it led to the creation of atomic bombs and, ultimately, nuclear weapons capable of destroying humanity.

Now consider a dual use dilemma that arose in the chemical industry in relation to the development of pesticides.

Option 1: Scientists morally ought to develop highly toxic pesticides since these enable the eradication of pests which destroy crops.
Option 2: Scientists morally ought *not* to develop these pesticides since they enabled the development of the nerve agent sarin which can be used by terrorists (e.g. Aum Shinrikyo in Tokyo in 1995) to kill innocent citizens and by rogue nation states to wage war in morally unacceptable ways (e.g. by Iraq in the Iraq/Iran war). NB: Sarin produces uncontrollable nerve cell excitation and muscle contraction leading to death by suffocation.

A potential dual use dilemma in the development of cyber-technology pertains to autonomous weaponised robots: weaponised robots that once programmed can select their targets and determine if and when the 'trigger' is to be pulled.

Option 1: Scientists morally ought to develop autonomous robots since these enable driverless cars and other beneficial technology.
Option 2: Scientists morally ought *not* to develop autonomous robots since they enable the development of autonomous weapons that reduce human control over the killing of humans.

Finally, consider a paradigmatic case of dual use research in the biological sciences, namely, the biological research on the deadly flu virus H5N1 which causes bird flu.

Scientists in the US and the Netherlands created a highly transmissible form of this deadly virus. Crucially, the work was done on ferrets which are considered a very good model for predicting the likely effects on humans. Accordingly, the dangers are very great indeed.

Option 1: Scientists morally ought to conduct research on the bird flu virus and do so intending to develop vaccines against similar naturally occurring and artificially created strains of H5N1.

Option 2: Scientists morally ought *not* to conduct the research since it will lead to the creation of a virus which is both highly virulent and easily transmissible to humans, and lead to the consequent far from remote possibility of the death of millions of humans—as the result, say, of a terrorist group launching a biological terrorist attack or of release of the virus into a human population due to negligence on the part of those working at a laboratory housing the virus.

In such cases the researchers—by going ahead with the research, and/or publishing their findings—will have foreseeably provided the means for the harmful actions of others and, thereby, arguably violated a moral principle. The principle in question is the principle of what we might refer to as the No Means to Harm (NMH) principle.[13] Roughly speaking, this is the principle not to provide malevolent persons with the means to harm; a principle which itself ultimately derives from the more basic principle: Do no harm.

   NMH is the principle that a single person or multiple persons should not avoidably and foreseeably (whether intentionally or unintentionally) provide others (directly or indirectly) with the means to intentionally (or negligently) do harm and it assumes: (i) the means in question is a means to do harm (including by virtue of being inherently dangerous material); (ii) there is a reasonable chance that the others in question (individually, in aggregate or by acting jointly, e.g. members of a terrorist group engaging in the manufacture of a so-called 'dirty bomb') will do harm; (iii) the harm in question is of very great magnitude (i.e. is very serious and on a very large scale), and; (iv) if there are multiple persons and/or the provision of the means by these multiple persons to others would be indirect, nevertheless, these multiple persons could and should see to it (if necessary by cooperating with one another) that the means in question is not provided.

   As with most, if not all, moral principles, NMH is not an absolute principle and, therefore, it can be overridden under certain circumstances. For example, it is morally permissible to provide guns to the police in order that they can defend themselves and others. Here there is an implicit invocation of the principle of necessity; it may be necessary for the police to possess guns. Likewise the principle of necessity may bear on the application of NMH in relation to dual use research. If there is no need, i.e. necessity, to conduct dual use research then, other things being equal, it should not be undertaken. If, for example, a piece of research can be undertaken to achieve some beneficial purpose which has little or no potential to be used for harmful purposes

---

[13]This principle, or similar ones, are familiar in a variety of ethical contexts. See, for example, Miller (2013) and Scanlon (1977).

then such research should be preferred to a piece of dual use research conducted for the same beneficial purpose. Indeed, other things being equal, the dual use research should not be undertaken.

A further principle that may bear on the application of the NMH in relation to dual use R&D (and the dissemination thereof) is the principle of proportionality. Consider, first, our policing example. Perhaps in some situations, such as riots, police may need to use CS gas with the consequence that not only rioters, but also some innocent bystanders may be harmed. Use of such methods should not be disproportionate, either in relation to the intended harming of rioters (consisting of intentionally causing them to inhale the gas fumes) or in relation to the risk of harm to innocent bystanders. Likewise, issues of proportionality may arise in relation to the application of NMH, albeit NMH is a principle of refraining from harming. Thus the potential harms consequent upon some kind of dual use R&D might be disproportionately large relative to the potential benefits.

Moreover, as is the case with many moral principles, in the application of NMH—including with respect to the involvement of the principles of necessity and proportionality—there is a degree of indeterminacy. If there are no currently available alternatives to dual use research on nuclear R&D for purposes of generating electricity, are there likely to be alternatives in the future, i.e. is this research really necessary? How much harm constitutes harm of very great magnitude? The loss of 100,000 lives obviously does, but what of the cyber-theft of 100,000 credit card numbers? Further, the application of the principle of NHM in the cases of interest to us is likely to involve multiple original researchers and multiple secondary users who are connected via complex indirect causal chains involving still others, e.g. those who communicate the research findings or provide hard to acquire materials to the secondary users. Finally, as is the case with the application of most, if not all, moral principles in complex situations involving multiple actors, the application of NMH is very often a matter of morally and empirically informed judgment. In the case of NMH the need for judgment depends in large part on the uncertainty of future harms.

2.4   What Dilemma and for Whom?

The dual use dilemma is a dilemma for scientists and technologists, but not only for scientists and technologists. For although the dual use phenomenon undoubtedly raises crucial moral or ethical (I use the terms interchangeably) questions about the duties and responsibilities of individual scientists and technologists, it is also an ethical issue for others. The phenomenon of dual use research and technology calls for important ethical decision making by actors (with duties and responsibilities) at many levels. Research institutions ought to decide how to oversee activities taking place within their confines and/or whether or not to provide (and perhaps require) educational programs on dual use issues. Scientific associations need to decide whether or not and/or how to address dual use research in codes of conduct; and they must decide whether or not and/or how to enforce such codes on members. Publishers need to decide what to publish and/or what screening mechanisms to put into place. Governments must decide whether or not and/or how to impose restrictions on dual use research and technology or, for that matter, whether to relax existing restrictions, e.g. in the case of research in the nuclear sciences undertaken for peaceful purposes. Governmental regulations could, among other things, potentially call for mandatory reporting of dual use research to committees for clearance before experiments are conducted or published and/or mandatory education of researchers about the dual use phenomenon and relevant ethical considerations. Funders of scientific research and technological development, finally, must decide what research and development to fund. Thus funders must decide whether or not relevant education, adherence to codes of conduct, and/or reporting of dual use research to committees before experiments are conducted or published should be conditions of individual researchers' or research institutions' eligibility for funding. The dual use phenomenon raises ethical issues for decision makers at each of these levels, because they all face the ethical question about how to strike a balance between security concerns and the promotion of academic freedom and/or scientific progress and technological development (assuming these things will sometimes come into conflict[14]). More detailed ethical analysis of the responsibilities of these other actors is therefore important.

Although government regulation of research and of technological development is controversial in some areas, such as the biological sciences (albeit not in others, such as the nuclear sciences) it may be imprudent to rely too heavily on voluntary participation by, or self-regulation of, scientists, technologists or the scientific community more generally. As we saw above in the discussion of WMDs, one reason is that scientists and technologists have participated in WMD R&D programs, including those of authoritarian regimes. (Arguably, other things being equal, authoritarian regimes are more likely to use WMDs offensively, i.e., other than in the context of a war of self-defence, than are liberal democracies.) Another reason that mandatory measures might be called for is that scientists may not always have sufficient expertise to judge the security dangers that might result from their research and/or publications. Responsible decision making requires assessment of the security risks and social benefits likely to arise from any given experiment or publication. Scientists, however, usually lack training in security studies and thus have no special expertise for assessing security risks in particular. In some cases they are systematically denied access to information crucial to risk assessment.

Consider, for example, the case of the mousepox study conducted in 2001 in Australia. The research was undertaken to develop an infectious contraceptive for mice in order to control them and, thereby, protect crops. However, the effect was to create a highly virulent strain of mousepox with the implication that a similar process might increase the virulence of smallpox. Accordingly, a primary concern was the

---

[14]Some might argue that free/open science would provide the best means to maximization of security. It is not clear what the evidence for this proposition is. Would the world be more safe if, for instance, scientific know-how in relation to nuclear weapons technology was entirely free/open? I return to this issue in Chap. 3.

possibility of proliferation of smallpox from former Soviet weapons stockpiles of the virus—i.e., because bioweaponeers would need access to the smallpox virus in order to apply the mousepox genetic engineering technique to it in the hope of thereby producing a vaccine-resistant strain of smallpox. Any detailed information about smallpox proliferation, however, is classified information to which the vast majority of scientists would not have access. In this important case—which has been a paradigm example of dual use research of concern—ordinary scientists would thus be unable to make an informed assessment of the risks of publication.

A further reason not to rely too heavily on voluntary participation or self-regulations is that conflicts of interest and, more generally, collective action problems may often come into play (see Chap. 4, Sect. 4.4). For instance, given that career advancement in science is largely determined by publication record, a researcher may often have self-interested reasons for publishing potentially dangerous findings (even when this might not be in society's best interests, all things considered).[15]

Yet another reason why ethical analysis of dual use research should not focus exclusively on the social responsibilities of scientists is that their duties (regarding whether or not to pursue a particular path of research or publish a particular finding) cannot be determined in a vacuum. What exactly an individual should or should not do partly depends on actions taken by other actors at other levels in the science governance hierarchy and, for that matter, in the military institutions of which science laboratories and the like might be a part.

Given the ultimate aim to avoid the malevolent (or otherwise culpable) use of dual-use technologies, it is important to recognize various stages in the "dual use pipeline" where preventative activities might take place—or where regulations might operate. First, there is the conduct of research that leads to dual-use discoveries. One way to prevent malevolent use is thus to prevent the most worrisome experiments from taking place to begin with. A second way to prevent malevolent use would be to prevent dissemination of dangerous discoveries after they are made—i.e. by not publishing them oneself (self-censorship), or by stopping others from publishing them (censorship). A third way would be to prevent malevolent use by limiting who has access to dual use technologies and materials such as "select agents" or potentially dangerous DNA sequences, requiring licensing of those using such technologies/materials, registration of relevant equipment, and so forth. A fourth way would be to strengthen the various conventions and treaties, such as the Nuclear Non-Proliferation Treaty and the Biological and Toxin Weapons Convention (BTWC)—in the latter case via the addition of verification methods or other such measures. This would help prevent state actors, at least, from using legitimate science for the promotion of offensive nuclear, biological, chemical and radiological weapons programs.

The point here is simply that the question of whether or not a researcher or technologist has a duty to refrain from a particular research and development project or publish a particular study partly depends on what preventative mechanisms have been put into place further down the "dual use pipeline". In the case of the chemical industry, considerable progress has been made in respect of regulation and, in

---

[15]Selgelid (2007).

particular, the implementation of the Chemical Weapons Convention. In the nuclear industry for historical reasons, and because of the high level of expertise and funding required, the levels of security are, speaking generally, quite high and, as a consequence, the possibility of terrorist groups or malevolent 'lone wolves' developing nuclear weapons relatively low. However, the biological sciences are somewhat different. If one discovers how to synthesize an especially contagious and/or virulent pathogen for example, the propriety of publishing this partly depends on whether regulatory measures that would prevent this finding from being employed by malevolent actors have been implemented. If there were stronger controls over access to the technologies and materials (e.g., DNA sequences) required by others to reproduce such a pathogen and/or if the BTWC was strengthened by addition of verification measures, for example, then the dangers of malevolent use arising from such a publication would be lower than would otherwise be the case. Thus whether or not a researcher would have a duty not to publish in such a scenario (assuming they were at liberty to do so) depends, at least partly, on whether or not policy makers et al. have fulfilled their duties to design adequate preventive measures and put them in place.

2.5   Ethical and Regulatory Dual Use Issues

My primary concern in this work is with moral or ethical principles and values, as opposed to detailed legal or, for that matter, regulatory rules. There is, of course, a close relationship between the moral and the legal and, more specifically, between the moral and the regulatory. For instance, typically criminal laws, such as the laws against murder, assault and theft, 'track' or follow antecedent moral principles; there is a law against murder, for example, precisely because we regard murder as *morally* wrong. Nevertheless, the moral and the legal are conceptually distinct, and the distinction needs to be kept in mind in what follows. An important corollary of the existence of this moral/legal distinction is that it is not necessarily the case that every research practice rightly regarded as immoral or unethical should always be made unlawful. There is also a close relationship between the legal and the regulatory. Thus many activities that are in themselves lawful are, nevertheless, subject to regulation in the manner in which they are conducted, e.g. health and safety regulations governing food production (food production per se being legal). However, regulation is not necessarily government regulation; professional activity, for example, is typically subject to regulations devised and imposed by professional associations.

There are a number of general moral and regulatory issues that need to be addressed in relation to dual use issues in R&D in science and technology. They include the following ones.

(A) Morally and Legally Impermissible Research and Development

- What, if any, research in the chemical, nuclear, cyber and biological sciences that gives rise to a dual-use dilemma is completely morally unacceptable and, therefore, ought to be unlawful?
- What purposes are served by dual-use research and development; specifically, what harms and benefits are consequent (or likely to be consequent), upon this research and development, e.g. increasing human understanding, saving lives, generating profits, enabling new methods of warfare?
- Assuming that it is the national and international legislators (and their respective communities) who ought to decide what general kinds of dual use research and development, if any, ought to be unlawful by virtue of the grave risks that they pose, who is to decide what specific research and development is in fact of one or more of the kinds in question, e.g. government committees, university-based biosafety/biosecurity committees, members of international bodies?

(B) Physical and Legal/Regulatory Conditions under which (Permissible) Experiments of Concern ought to be undertaken

- Who (personnel) or what (organisations) ought to be allowed to undertake dual use research?
- In relation to the various categories of *prima facie* permissible research that, nevertheless, give rise to dual-use dilemmas, what are the safety and security—and associated regulatory—conditions under which this research ought to be undertaken, e.g. background checks and security clearance for research personnel, training programs, licensing of laboratories?

(C) Development of Dual-Use Technology

- Who (personnel) or what (organisations) ought to be allowed to develop dual use technologies?
- What amounts of what materials, e.g. chemical stockpiles, ought to be allowed and for what purposes?
- In relation to the various categories of *prima facie* permissible development of technologies that, nevertheless, give rise to dual-use dilemmas, what are the safety and security—and associated regulatory—conditions under which this development ought to be undertaken, e.g. background checks and security clearance for research personnel, training programs, licensing of organisations?

(D) Commercialisation of Dual Use Research

- What dual use research ought to be allowed to be commercialised, i.e. undertaken in the private sector for profit?
- Under what conditions should commercialised dual use research be allowed, e.g. screening of buyers, licensing of sellers.

(E) Dissemination

- In relation to permissible, safe and secure research in the chemical, nuclear, cyber and biological sciences that, nevertheless, gives rise to dual-use dilemmas what, if any, restrictions ought to be placed on its dissemination in scientific journals, the mass media etc.?
- In relation to permissible, safe and secure research in these sciences that, nevertheless, gives rise to dual-use dilemmas who ought to decide what, if any, research findings ought not to be disseminated or ought to have restrictions placed on their dissemination, e.g. journal editors, newspaper editors?

(F) Regulatory Authorities

- What regulatory architecture ought to be put in place internationally, nationally and at the industry-wide and individual organisational level in universities and in commercial firms to address the ethical concerns with, and to support the regulation of dual use research/dissemination in the chemical, nuclear, cyber and biological sciences?
- What regulatory authorities ought to be established to advise international bodies, governments and others on dual use issues and to ensure compliance with regulations?

2.6 Conclusion

In this chapter I have provided a definition of the concept of dual use in relation to chemical, nuclear, cyber and biological R&D. On this (somewhat stipulative) definition new or emerging science or technology is dual use if:

(1) It can be used for both beneficial and harmful purposes—where either the harmful purposes involve the use of weapons as means, and usually WMDs in particular, or the serious, large-scale harm aimed at does not necessarily involve weapons or weaponisation;
(2) The serious, large-scale harm in question is caused by a single act of using the technology—as opposed to multiple acts that in aggregate cause great harm;
(3) A large-scale beneficial outcome is intended by the original researchers;
(4) The actual or potential harmful outcome is reasonably foreseeable by the original researchers and, if it eventuates, is either intended by secondary malevolent users or, at least, their secondary use involves culpable negligence.

I have also introduced the No Means to Harm principle (NMH) as central to the moral responsibilities of scientists and technologists and outlined in general terms the ethical and regulatory issues requiring attention if dual use issues are to be satisfactorily resolved.

References

Meier, Oliver, and Iris Hunger. 2014. *Between Control and Cooperation: Dual-use, Technology Transfers and the Non-Proliferation of Weapons of Mass Destruction*, 17. Osnabruck: DSF.

Miller, Seumas. 2013. Moral Responsibility, Collective Action Problems and the Dual Use Dilemma in Science and Technology. In *On the Dual Uses of Science and Ethics*, ed. Michael Rappert, and Brian Selgelid. Canberra: ANU Press.

Miller, Seumas, and Michael Selgelid. 2007. Ethical and Philosophical Consideration of the Dual Use Dilemma in the Biological Sciences. *Science and Engineering Ethics* 13: 523–580.

National Research Council. 2004. *Biotechnology Research in an Age of Terrorism*. Washington, D.C.: National Academies Press.

National Science Advisory Board for Biosecurity. 2015. *Framework for Conducting Risk and Benefit Assessments of Gain-of-Function Research*. Washington, D.C.: NSABB.

Rappert, Brian, and Michael Selgelid (eds.). 2013. *On the Dual Uses of Science and Ethics*. Canberra: ANU Press.

Scanlon, Thomas. 1977. A Theory of Freedom of Expression. In *The Philosophy of Law*, ed. Ronald M. Dworkin. Oxford: University Press.

Selgelid, Michael. 2007. A Tale of Two Studies: Ethics, Bioterrorism, and the Censorship of Science. *Hastings Center Report* 37 (3): 35–43.

Selgelid, Michael. 2016. Gain of Function Research: Ethical Analysis. *Science and Engineering Ethics* 22 (4): 923–964.

Tucker, Jonathan (ed.). 2012. *Innovation, Dual Use, and Security: Managing the Risks of Emerging Biological and Chemical Technologies*. Harvard, MA: MIT Press.

US Department of Health and Human Services. 2004. Fact Sheet: *HHS Fact Sheet Project Bioshield*. http://www.hhs.gov/news/press/2004pres/20040721b.html. Accessed 27 June 2006.

van der Bruggen, Koos, Seumas Miller, and Michael Selgelid. 2011. *Report on Biosecurity and Dual Use Research*, 1–122. The Hague: Dutch Research Council.

Chapter 3
Collective Knowledge and Collective Ignorance

Abstract Scientific knowledge—a species of collective knowledge—contributes greatly to human well-being; yet scientific knowledge enables technologies that can be extremely harmful. Accordingly, the question arises as to whether we ought to aim at ignorance and, in particular collective ignorance, rather than scientific knowledge, of certain technologies. We might do this by means of banning certain scientific research, e.g. into biological weapons, and/or by censorship of certain scientific findings. In this chapter I provide a taxonomy of concepts of collective knowledge (e.g. public propositional knowledge, expert practical knowledge) and an account of the related concepts of collective ignorance. In doing so my concern is with the concepts of collective knowledge and ignorance relevant to harmful technology and, especially, scientific knowledge/ignorance in the chemical industry, nuclear sciences, cyber-technology field and biological sciences relevant WMDs—such knowledge/ignorance being salient in discussions of dual use issues.

As we have seen, on the one hand, scientific knowledge contributes greatly to individual and collective well-being, e.g. enabling nuclear power stations producing low cost electricity; so evidently scientific knowledge is a good thing and ignorance of it a bad thing. On the other hand, scientific knowledge enables technologies that can be extremely harmful to individuals and collectives, e.g. nuclear weaponry. So, at least with respect to some technologies, evidently knowledge is a bad thing and ignorance a good thing. Accordingly, the question arises as to whether we ought to aim at ignorance, rather than scientific knowledge, of certain technologies and, therefore, curtail scientific research in these fields. If this is so then the question arises as to which technologies (and which scientific research programs).[1] For instance, perhaps research into enhancing the transmissibility into humans of highly virulent pathogens ought to be curtailed. After all, it is widely believed that R&D into biological weapons ought not to be conducted.

In this chapter in Sect. 3.1 I provide a taxonomy of concepts of collective knowledge and in Sect. 3.2 an account of the related concepts of collective ignorance. In

---

[1] An earlier and more detailed version of the material in Sects. 3.1 and 3.2 of this chapter appeared in Miller (2017).

© The Author(s) 2018
S. Miller, *Dual Use Science and Technology, Ethics and Weapons of Mass Destruction*, SpringerBriefs in Ethics, https://doi.org/10.1007/978-3-319-92606-3_3

doing so my concern is with the concepts of collective knowledge and collective igno-rance relevant to harmful technology and, especially, scientific knowledge/ignorance in the chemical industry, nuclear sciences, cyber-technology field and biological sci-ences relevant WMDs—such knowledge/ignorance being salient in discussions of dual use issues. The implications of these notions of collective knowledge and col-lective ignorance for dual use issues are explicitly discussed in Sect. 3.3.

3.1 Collective Knowledge

Our starting point is the invocation of a familiar threefold distinction made in respect of individual (as opposed to collective) knowledge. Firstly, there is *knowledge-by-acquaintance*: knowing someone or something.[2] For example, if two strangers have a face-to-face conversation then there is direct (physical and psychological) expe-rience of one another; there is, therefore, knowledge-by-acquaintance. For ease of exposition I sometimes refer to this kind of knowledge as acquaintance-knowledge (whether in its individual or collective form).

Secondly, there is *propositional knowledge*: knowledge of the truth of some propo-sition.[3] This is knowledge that, for example, some state of affairs obtains. Proposi-tional knowledge is expressed in language by sentences with a subject and a predicate. Consider a detective who knows that the fingerprints found on a knife at a particu-lar crime scene were those of the suspect. Here there is trace material found at the crime-scene, namely, the fingerprints on the knife, and this trace has been caused by the suspect handling said knife. The detective has propositional knowledge of this state of affairs if he or she knows it to be the case and has expressed this knowledge in a sentence(s) of a language.

Note that whereas propositional knowledge is expressed in language, it is not necessarily expressed in a form accessible to others; it might remain in the realm of inner thought. Thus the detective might know that Jones is the murderer and express this thought to himself in a sentence, but the detective does not necessarily utter this sentence for others to hear it; he does not necessarily assert out loud or make a written statement expressing his propositional knowledge.

Thirdly, there is *knowing-how*.[4] To know how to do something (e.g., knowing how to ride a bike, knowing how to read an x-ray film), is in essence to possess a skill. Knowledge-by-acquaintance and propositional knowledge are cognitive states whereas knowing-how is essentially practical in character and, as such, more closely aligned with conative rather than cognitive states. For ease of exposition I sometimes refer to this kind of knowledge as practical knowledge (whether in its individual or collective form).

---

[2]A relate distinction was made famous by Bertrand Russell. See Russell (1910, 108–128).

[3]For a useful introduction see Lehrer (1990).

[4]See, for example, Polanyi (1967) and Hetherington (2011).

Clearly scientists need to have all three sorts of knowledge. They need to verify certain claims by direct observation (acquaintance-knowledge). They also have to have, and be able to obtain and communicate, propositional-knowledge, e.g. in scientific publications and verbal communication with one another. In addition, they need to know how to do various things, e.g. use a microscope, (practical-knowledge).

Moreover, these three different types of knowledge are *interdependent*. Practical-knowledge, (e.g. how to use a microscope) typically depends on acquaintance-knowledge (e.g. seeing and grasping the microscope). And the methods of acquiring new propositional-knowledge often depend on acquaintance-knowledge (e.g. observation), and practical-knowledge, (e.g. how to use scientific equipment), as do the latter two types on propositional knowledge (e.g. a written manual describing scientific equipment and how to use it).

What of collective knowledge?[5] The salient notions of collective knowledge in the philosophical literature tend to be species of propositional knowledge. These are often referred to as common knowledge, mutual knowledge, mutual true belief and the like.[6] These notions are typically constructed out of the notion of mutual true belief. Thus two agents, A and B, mutually believe truly that p if A believes truly that p, B believes truly p, A believes truly that B believes truly p, B believes truly that A believes truly that p, and so on. Note that if one agent has beliefs with respect to another agent's beliefs and vice versa, in this manner, I will say that their beliefs are *interconnected*.[7]

Mutual knowledge—in the sense of mutual true belief—is closely related to another concept, namely, that which I will refer to as openness.[8] Openness is the social or interpersonal analogue of knowledge-by-acquaintance and, as such, is not necessarily propositional in character. For openness is mutual sensory awareness (hereafter mutual awareness) of an object and of oneself and the other person(s) as having awareness of that object. In the case of linguistic 'objects', speakers and hearers have mutual sensory awareness of utterances of sentences, i.e. of certain sorts of structured sounds and marks. Perhaps openness entails mutual true belief, but the reverse is not true; there can be mutual true belief without openness. For example, two people in a room could have mutual true beliefs with respect to an unseen, unheard etc. object in an adjoining room.

If openness is the social or interpersonal analogue of individual acquaintance-knowledge, *joint* knowing-how is the social or interpersonal analogue of individual practical-knowledge. Joint knowing-how finds expression in joint action; joint actions are the exercise of joint knowledge-how.

The notion of joint action per se is a familiar one in the philosophical literature.[9] Roughly speaking, joint actions are actions involving a number of agents performing

---

[5]Schmitt (1994), Goldman (1999), Kusch (2002).

[6]For convenience, we use the term "mutual" rather than "common" when referring to the kind of phenomena in question. For definitions of some of these notions see, for example, Smith (1982).

[7]Similarly for like mental states, including states of awareness.

[8]Miller (2015).

[9]See, for example, Miller (1992, 275–299) and Miller (2001).

interdependent actions in order to realise some common goal. Examples of joint action are a number of tradesmen building a house and a team of researchers seeking the cure for cancer. Joint actions are interdependent actions directed toward a common goal or end.

What of joint knowing-how? Consider the joint task of rowing a boat or dancing with a partner. One partner might know how to dance the tango, for example, but the other might not. If so, the two partners will not be able to dance the tango together. On the other hand, if both know how to dance the tango, i.e. how to perform their respective dance roles as lead and follower, then it is likely they *jointly* know how to dance the tango. Accordingly, they can proceed to exercise their joint know-how by performing the joint action of dancing the tango.

Since one or more persons could have exhaustive mutual propositional knowledge concerning dancing the tango but yet not know how to dance the tango, it appears that joint knowing-how is not a species of collective propositional knowledge. In short, collective practical-knowledge is not a species of collective propositional-knowledge.

Collective practical-knowledge is ubiquitous, at least in modern societies. Consider the building of a skyscraper building. This involves architects, engineers, bricklayers, carpenters, electricians etc., all of whom have specific forms of individual practical-knowledge (individual know-how, so to speak), but none of whom are individually possessed of all the different forms of practical-knowledge. Accordingly, their *collective* practical-knowledge (*joint* know-how) is required in order to realise the collective end of constructing the skyscraper. The same point holds for the designing and construction of nuclear facilities/technology, chemical plants/technology etc.

Thus far we have distinguished three forms of collective knowledge, namely, propositional (mutual knowledge), acquaintance (mutual awareness), and practical (joint knowledge-how). However, there are two additional species (or, perhaps, subspecies) of collective knowledge that need to be identified. The first of these we will refer to as public-knowledge, the second as expert-knowledge. These two species of collective knowledge have a propositional and a practical form.[10]

In its propositional form public-knowledge consists of true propositions that are matters of individual knowledge in the ordinary sense for some persons, i.e. it is 'in their heads', but for many or most these propositions are only knowledge in the sense that they are available for acquisition.[11] Thus much of the information stored in hardcopy format in books in libraries, in softcopy format in electronic data-bases, in public records, (e.g., court records) is public-*propositional*-knowledge. Again, the propositional knowledge in the heads of relevant public officials, such as those serving in information counters at railway stations, is public knowledge in our sense.

In its practical form, public-knowledge consists of individual know-how (e.g. how to bake a cake, how to drive a car, how to read and write) that is either actually possessed, or is available for acquisition, by all or most members of some 'public'. Thus the widespread availability of 'how to' manuals, driving lessons, primary school edu-

---

[10]They may well also have a knowledge-by-acquaintance form but, if so, this is not central to our concerns in this chapter so I omit discussion of it.

[11]Popper (1972, Chap. 4).

cation and, in the end of the day, the widespread access to human persons possessed of the relevant 'know-how' and capable of inducting others into it, ensures that there is public-*practical*-knowledge.

Expert-propositional-knowledge is knowledge 'in the heads' of the members of some group (the experts) in the form of mutual knowledge, but this knowledge is not 'in the heads' of another group (the non-experts). Expert-knowledge, like public-knowledge, is frequently stored in libraries, data-bases and so on that are, at least in theory, accessible to the public, i.e. the non-experts.[12] However, expert-knowledge is *not readily understandable* by ordinary members of the public, and so it is not in a substantive sense available to them. Thus much scientific knowledge in academic journals is expert-propositional-knowledge, but not public-propositional-knowledge.[13]

Expert-*practical*-knowledge is actually possessed by experts or is readily available to them, e.g. by way of professional top-up training courses. Expert-*practical*-knowledge is akin to expert-propositional-knowledge in that it is not in a substantive sense available to the public. For example, the surgeon's knowledge-how to perform open-heart surgery is limited to those who gain access to medical schools, pass examinations, and so on.[14]

Let us now summarise our taxonomy of collective knowledge. There are three basic forms of collective knowledge corresponding to individual propositional-knowledge, individual acquaintance-knowledge and individual practical-knowledge. The three basic forms are (respectively):

(1)   collective propositional-knowledge (mutual knowledge);
(2)   collective acquaintance-knowledge (mutual awareness);
(3)   collective practical-knowledge (joint knowledge-how).

Moreover, collective propositional-knowledge has two salient species for our purposes here, namely:

(1a)   public (propositional) knowledge and
(1b)   expert (propositional) knowledge,

as does collective practical-knowledge, namely:

(3a)   public (practical) knowledge and
(3b)   expert (practical) knowledge.

---

[12]In some cases, of course, this is not so, e.g. classified nuclear technological knowledge.

[13]This expert propositional knowledge often goes hand in glove with expert knowing-how. Consider, for example, a surgeon's propositional knowledge of aspects of surgery.

[14]It is, of course, true that the distinction between public knowledge and expert knowledge is not always clear-cut; in many domains of knowledge the one shades into the other, e.g. historical knowledge.

This gives a total of five types of collective knowledge of which two are species of propositional-knowledge, two are species of practical-knowledge and one is a species of acquaintance-knowledge.[15]

3.2 Collective Ignorance

In light of this above account of collective knowledge, what are we to make of the notion of collective ignorance?[16] It is tempting simply to define collective ignorance as the absence of collective knowledge. Since there are five types of collective knowledge there will be, on this account, five corresponding types of collective ignorance. However, this simple account is not adequate.[17]

Before proceeding further we need to invoke a distinction made with respect to individual ignorance, namely, between what I refer to as doxastic and non-doxastic ignorance.[18] The doxastic ignorance of person A with respect to the proposition p (where p might be either true or false) obtains only if A suspends judgement with respect to p. Typically in such cases, A *believes* that he does not know whether or not p. By contrast, non-doxastic ignorance of A with respect to p obtains only if A does *not* have any beliefs (or related doxastic attitudes) with respect to p (including higher order beliefs, such as the belief that he does not know whether p). Typically, in such cases, A has never contemplated whether or not p.

Note that doxastic ignorance has no clear analogues in cases of acquaintance-knowledge or practical-knowledge. There can, of course, be doxastic ignorance in the sense of a *belief* (or other doxastic state) that one is not aware of object O. But it is doubtful that one could be *aware* of one's unawareness of O since, arguably, one cannot be *aware* of 'something' that is a mere absence i.e. one's unawareness of O. Again there can, of course, be doxastic ignorance in the sense of a *belief* (or other doxastic state) that one does not know how to x, but surely the idea of A knowing how to not know how to x makes little sense. For such higher order know-how seems to presuppose the lower order know-how one knows how not to have.[19] At any rate, in what follows I set aside these putative (non-propositional) higher order forms of ignorance.

Note also that on this dualistic (doxastic/non-doxastic) account of ignorance, if A falsely believes that p then A is not ignorant of p, although A is wrong about p. Note

---

[15]Evidently, since acquaintance knowledge is generally available to everyone possessed of ordinary perceptual faculties and is not necessarily linguistic in form it generally does not have counterpart expert and public knowledge species.

[16]There is some philosophical literature on individual ignorance (Unger 1974) but little, if any, on collective ignorance.

[17]Nor is it adequate in respect of individual ignorance. See Peels (2010, 57–67).

[18]Here I utilize to some extent the work of Peels (2010),"What is Ignorance?"

[19]Of course, A might know-how to bring it about that A (or, indeed, B, C etc.) does not know-how to x, e.g. by destroying the relevant part of his brain that enables him to know how to x. But this is a different matter.

further that on this account if A does not have any justification for A's true belief that p then A is not ignorant, albeit one might want to hold that A does not have *knowledge* of p in some stronger sense of knowledge than true belief that p (since A lacks any justification for his belief that p).

Armed with the above account of collective knowledge and with this distinction between doxastic and non-doxastic ignorance, can we now define collective ignorance? Not quite. For before doing so we need to make one further distinction. This is the distinction between collective knowledge and aggregate knowledge; and, by parity of reasoning, between collective ignorance and aggregate ignorance.

Consider first a distinction between *collective* knowledge and *aggregate* knowledge. Let us first consider aggregate propositional-knowledge. If A knows that p, B knows that p, C knows that p etc., but neither A, nor B, nor C etc. has any beliefs with respect to the knowledge that p of the others, then there is no collective knowledge. Rather there is, what I refer to as, aggregate knowledge. There is aggregate, but not collective, knowledge since there is no interconnection (see Sect. 3.1 and discussion below) or interdependence (see Sect. 3.1 and discussion below) between the beliefs (and, therefore, knowledge) of the agents in question. There is no interconnection since, for instance, A does not believe that B believes that p. There is no interdependence because, for instance, A's belief that p is not dependent on B's belief that p. What of aggregate ignorance and collective ignorance?

A preliminary point to be made here concerns cases of aggregate (but not collective) *knowledge*. Are such cases necessarily cases of *collective ignorance*? After all, such cases are, ex hypothesi, not cases of collective knowledge and if ignorance is merely the absence of knowledge then, it might be suggested, aggregate knowledge (not being collective knowledge) must be collective ignorance. This suggestion should be rejected. For one thing, the idea that aggregated states of *knowledge* could constitute *ignorance*, even if collective ignorance, is somewhat paradoxical. For another thing, there is no interconnection or interdependence between these aggregated states of knowledge; they fail these collectivity tests.[20] Here, as mentioned above, a mental state of one agent (e.g. A believes that B believes that p), is *connected* in the relevant sense to another agent's mental state (e.g. B believes that p), if B's mental state figures in the content of A's mental state. There is *inter*connection if there is a two way connection between the mental states of two or more agents (e.g. [A believes that B believes that p] and [B believes that A believes that B believes that p]). By contrast, a mental state of one agent (e.g. A believes that p) is *dependent* in the relevant sense to another agent's mental state (e.g. B believes that p), if the latter is a (subjectively held) reason for the former (e.g. if A believes that p at least in part because B has communicated to A that B believes that p). There is *inter*dependence if there is two-way dependence [e.g. if B intends to communicate B's belief that p to A at least in part because A believes that B would not intentionally communicate what B believes is false (and, of course, B believes that A has this latter belief)].

---

[20]The same general point could be made of attempts to characterize mere aggregates of practical or acquaintance knowledge as instances of collective ignorance.

Let us, then, turn to a more obvious candidate for collective ignorance, namely, aggregate *ignorance*. Consider first aggregate *non-doxastic* ignorance. In such cases A has no belief (or other doxastic state) with respect to p, likewise B, C etc. Nor does A have any beliefs (or other doxastic state) with respect to B's (or C's etc.) beliefs (or lack thereof) with respect to p. What of the aggregates comprised of such absent 'states'? There is no interconnection (or interdependence[21]) between these absent 'states' of A, B, C etc. Accordingly, these cases also fail the interconnection (and interdependence) tests and, therefore, are not instances of *collective* ignorance.

I take it that the same general point can be made in respect of the analogous cases of aggregate non-doxastic acquaintance-ignorance and aggregate non-doxastic practical-ignorance (and, for that matter, analogous cases of aggregate non-doxastic public and expert ignorance, whether propositional or practical in form). Given that such cases do not involve any interconnection or interdependence they are not instances of collective ignorance, but are merely instances of aggregate ignorance. Thus, to take aggregate non-doxastic practical ignorance as an example, A does not know how to x, B does not know how to x, C does not know how to x etc., and, A has no belief (or other doxastic state) with respect to A, B, C etc. knowing how to x, likewise B, C etc. Nor does A have any beliefs (or other doxastic state) with respect to B's (or C's etc.) beliefs (or lack thereof) with respect to how to x. Accordingly, I set aside all forms of aggregate non-doxastic ignorance. None are forms of collective ignorance.[22]

Let us now consider aggregate *doxastic* ignorance. This form of ignorance involves cases in which, for example, A, B, C etc. each has a belief (indeed, a true belief[23]) that he or she does not know whether or not that p. It also involves cases of aggregate doxastic ignorance with respect to awareness (e.g. A, B, C etc. each has a true belief that he or she is not aware of O) and cases of aggregate doxastic ignorance of practical knowledge (e.g. A, B, C etc. each has a true belief that he or she does not know how to x.) As with the corresponding non-doxastic cases, these cases of doxastic ignorance being mere aggregates are not instances of *collective* ignorance. For they do not necessarily involve any interconnection or interdependence between their constitutive individual doxastic states. For instance, A's belief that he does not know whether or not p does not refer to B's belief that she does not know whether or not that p. Likewise for the corresponding beliefs of B and C etc.

---

[21]Matters might be different if A, B, C etc. had contrived somehow to jointly bring it about that each did not know that p and in a manner that did not involve any of them contemplating whether or not that p (or any higher order belief that p). This scenario seems extremely doubtful.

[22]Of course, if someone wants to insist that aggregate ignorance is a form of collective ignorance and, thereby, reject my interconnection and interdependence tests for collective ignorance (and collective knowledge) then we will have a verbal dispute about the meaning of "collective" but nothing of substance will follow.

[23]I assume that these beliefs are true ones in order to simplify matters. For example, if they were false beliefs then (contra the example) A, B, C etc. would know whether or not that p and thus these cases would not be cases of aggregate *ignorance*. On the other hand, if the beliefs in question were unspecified with respect to their truth or falsity then it is correspondingly indeterminate whether they should be regarded as cases of aggregate *ignorance*.

Accordingly, we should accept the general proposition that aggregate ignorance is not necessarily collective ignorance. More specifically, we should accept the following. If A, B, C etc. each individually truly believes that he or she does not know whether or not p (or that he or she is unaware of O or that he or she does not know how to x) and these true beliefs are not interconnected or interdependent then:

(i) A, B, C etc. have aggregate (doxastic) ignorance of p (or of O or with respect to how to x);
(ii) A, B, C etc. do *not* have aggregate *knowledge* of p (or of O or with respect to how to x);
(iii) A, B, C etc. do *not* have *collective* knowledge of p (or of O or with respect to how to x;
(iv) A, B, C etc. do *not* have *collective ignorance* of p (or of O or with respect to how to x).

And, to reiterate, the reason for (iv) is that their individual suspensions of judgement and resulting higher order true beliefs were neither interconnected nor interdependent. Accordingly, while A believes that A does not know whether or not p, B believes that B does not know whether or not that p and so on for C, etc., nevertheless, A does not have any beliefs with respect to B's, or C's etc. beliefs about p, nor does B have any beliefs with respect to A's, C's etc. beliefs about p; and similarly for C etc. Moreover, neither A, nor B nor C etc. suspended his or her judgement interdependently with the others doing so.

By way of contrast, consider an example in which there is both epistemic *interconnection* and *interdependence*. Assume some, but not all, of the members of a team of detectives individually fail to perform successfully their important contributory epistemic tasks in a murder investigation, e.g. A's forensic analysis is incorrect. As a consequence, there is mutual knowledge among the members of the team that: (1) they have jointly failed to come to know the identity of the murderer and; (2) each is individually ignorant of the identity of the murderer. Accordingly, there is interconnectedness of (true) beliefs among the detectives, e.g. A knows that B does not know who the murderer is. Moreover, there is epistemic interdependence among the detectives, e.g. B does not know who the murderer is because (in part) A's forensic analysis was incorrect.[24]

In this detective scenario since there is mutual knowledge that each does not know whether or not that p then, arguably, there is collective *ignorance*. For there is interconnection between the doxastic ignorance of each; it is not merely a case of aggregate (doxastic) ignorance. However, notice that the notion of knowledge, specifically, mutual knowledge (in the sense of mutual true belief), is required to differentiate aggregate ignorance from collective ignorance and, moreover, that mutual knowledge is a necessary component of collective (doxastic) ignorance.

I conclude, firstly, that there is no such thing as collective *non-doxastic* ignorance, but rather only collective *doxastic* ignorance and that, secondly, collective (doxastic) ignorance is a form of mutual knowledge, albeit mutual knowledge of ignorance.

---

[24]Miller and Gordon (2014, Chap. 2).

Moreover, there are different species of collective ignorance (i.e. of collective doxastic ignorance). However, prior to identifying these, we need to introduce the notions of a *molecule* of knowledge and a *web* of knowledge.

A *molecule of knowledge* is a unitary composite of propositional, acquaintance and practical knowledge, and each such molecule exists in its entirety 'in the head' of an individual person, albeit different token molecules of the same type can exist in other individuals. For example, agent A might have the molecule consisting of the propositional knowledge that John drives a Ferrari, acquaintance knowledge of John and of Ferraris, and practical knowledge of how to drive.[25]

A *web of knowledge* is an inferentially integrated cluster of molecules of knowledge. Moreover, a web of knowledge might exist in its entirety 'in the head' of an individual person. This is perhaps especially likely in the case of an expert in a discrete field of practical knowledge, such as knowledge of the internal combustion engine. However, it might not exist in its entirety 'in the head' of an individual person yet still exist in its entirety 'in the heads' of a set of individuals. If the web in question does not exist 'in the head' of one but only 'in the heads' of many, then each fragment of the web exists 'in the head' of some individual member of the relevant set of individuals. Moreover, each of these individuals knows that the set of individuals of which he is a member has a web of knowledge—and each can identify this web under more or less the same description—and each knows of his or her fragment of knowledge that it is a fragment of this web. So there is *joint* knowledge of the web, notwithstanding that each only has detailed knowledge of his or her fragment[26] and each might, in fact, be quite ignorant of the details of the other fragments. Accordingly, there is the possibility of individually or jointly acting on the basis of the web. For example, a web of knowledge might consist of the knowledge that John drives a Ferrari in London (understood as a molecule since John has beliefs about his Ferrari, is sensorily acquainted with his Ferrari and knows how to drive it), Fred rides a bicycle in London (a second molecule), Mary uses the London Underground (a third molecule) and, therefore, there are at least three different modes of transport in London (inferentially derived molecule). So even if Fred has never seen, knows little about and does not know how to drive a Ferrari, and has never been to, knows little about and does not how to use the Underground, nevertheless, he is aware that cars and underground trains are available forms of transport in London and he might, for example, seek out Mary to show him how to use the Underground. For our purposes here, a more relevant example of a web of knowledge would be the knowledge required to build a nuclear weapon or to weaponise a virulent biological agent (a pathogen) or toxic chemical. In these examples, it is conceivable that no single scientist has the entirety of the web of knowledge in his or her head but rather only a fragment thereof. An example of a molecule of such a web of knowledge might be a bench scientist's detailed knowledge of anthrax (since the molecule of

---

[25]I assume that one can have practical knowledge without exercising it at a given moment and, therefore, it is 'in one's head' in this sense.

[26]Or, at least, adequate knowledge of his or her fragment relative to the requirement for joint knowledge of the web of knowledge in question.

knowledge is, let us assume, in the head of the single scientist). However, the scientist in question might not know how to weaponise anthrax, although she might know that others know how to weaponise it. Likewise those who know how to weaponise anthrax might not have the detailed knowledge of anthrax possessed by the bench scientist and might, therefore, need to rely on the bench scientist if the process of weaponisation is to be successfully realised.

In the light of the above, let us now identify three salient senses of our above notion of collective ignorance.

(1) There is mutual knowledge among A, B, C etc. that each does not have any molecule member of a given set of molecules of knowledge (i.e. a structured set that could potentially be a web of knowledge, W).
(2) There is mutual knowledge among A, B, C etc. that each does not have a given web of knowledge, W, (comprised of the structured set of molecules mentioned in (1) above).
(3) There is mutual knowledge among A, B, C etc. that they do not jointly have the web of knowledge, W, (mentioned in (2) above).

I take it that (3) is of greatest interest to us, in the context of our focus on dual use science and technology, although (2) is not without interest as will emerge below. There is, however, a residual matter, namely, collective ignorance in respect of public knowledge and/or of expert knowledge. The above account of collective ignorance can be adjusted to accommodate collective *public* ignorance and collective *expert* ignorance (in both their propositional and practical forms). The result is the following bifurcated definition of collective ignorance in sense (3) in the case of a group comprised of experts and non-experts.

Collective Public Ignorance: Members of some group, G, comprised of experts and non-experts, have collective *public* ignorance of web of knowledge, W, if and only if: there is mutual knowledge among members of G as a whole that they do not jointly have W—even if members of the sub-group of experts jointly have W—and that they cannot readily come to jointly have W by accessing available knowledge storage centres or knowledgeable persons, such as members of their expert sub-group.

Collective Expert Ignorance: Members of some group, G, comprised of experts and non-experts, have collective *expert* ignorance of web of *expert* knowledge, W, if and only if there is mutual knowledge among members of G (or, at least, among the members of the expert sub-group) that neither the members of G as a whole, nor even the members of the expert sub-group, jointly have W and that neither the members of G as a whole, nor even the members of the expert sub-group, can readily come to jointly have W by accessing available knowledge storage centres or knowledgeable persons from outside G.

3.3   Collective Knowledge, Collective Ignorance and Dual Use Technology

Scientific and technological knowledge is comprised in part of the propositional, acquaintance and practical knowledge of individual scientists and engineers. However, this knowledge is not merely aggregate knowledge, it is also collective knowledge (in all five senses of collective knowledge adumbrated above). Indeed, typically, it consists of molecules of knowledge and comprises a web, or webs, of joint knowledge. Moreover, much of this collective knowledge is morally significant; certainly the collective knowledge with respect to dual use science and technology is morally significant. Given its collective character and its moral significance, a question arises with respect to collective moral responsibility for acquiring such collective knowledge and, potentially, for refraining from acquiring it (or, at least, from disseminating it in a manner that enables it to become public, as opposed to expert, knowledge). Here we need a serviceable account of collective moral responsibility.[27] This will be provided in detail in Chap. 4. According to this account, collective moral responsibility is a species of individual responsibility; specifically, joint responsibility. That is (roughly speaking), each individual member of a group has a moral responsibility *jointly held with the other members*.

Research and development of WMDs is constituted in large part by collective (scientific and engineering) knowledge—collective knowledge of webs of knowledge. In the case of nuclear technology, the webs of knowledge in question are jointly possessed—but not possessed in their entirety by any single individual—and they are possessed only by experts. Evidently, no single individual scientist or engineer, and no ordinary member of the public, could successfully research and develop a nuclear weapon acting alone.

Surely scientists and engineers have a pro tanto collective moral responsibility to refrain from the research and development of WMDs, given the lethal threat that WMDs pose to human-kind. If so, this collective moral responsibility might trump, for instance, their collective institutional responsibility as members of a nation-state's defence force to engage in research and development of WMDs. Moreover, since research and development in WMDs is constituted in large part by collective expert (scientific and engineering) knowledge (a jointly held web of knowledge), arguably, they also have a collective moral responsibility to maintain or bring about a state of collective ignorance with respect to such R&D, i.e. among all nation-states and other groups. In short, they have a collective moral responsibility to bring it about that there is mutual knowledge among them and others (e.g. members of governments, members of the various publics) that they and others (e.g. future scientists) do not jointly have the web of knowledge in question.

Of course, it might be argued that since the webs of knowledge in question already in large part exist, this is an impossible task and, therefore, it cannot be a matter of moral responsibility, collective or otherwise. Against this it might in turn be argued

---

[27]Miller (2006, 176–193). See also Miller (2010, Chap. 4).

that, at least in the case of nuclear technology, much could be done short of securing complete collective ignorance. For example, the dissemination of the collective knowledge in question could be curtailed (as is already the case, to a considerable degree by way of being 'classified') and this knowledge restricted to scientists and engineers functioning in 'responsible' nation-states. Consider, in this connection, the recent Iranian nuclear arms technology deal orchestrated by President Obama.

Whatever the possibilities of collective ignorance in respect of nuclear arms technology, apparently matters are even more problematic in the case of the research and development of biological weapons. For it might be possible in the not too distant future for someone with only rudimentary scientific and engineering knowledge to weaponise a biological agent in their garage. If so, the form of collective ignorance required is (2) above, namely, that there is mutual knowledge among A, B, C etc. that *each* does not have the web of knowledge in question. The putative collective responsibility to bring about this state of collective ignorance is surely an onerous one, arguably, impossibly onerous. We return to this issue in Chap. 8.

Let us assume that the relevant scientists and engineers have at least some collective responsibilities with respect to some species and/or degrees of collective ignorance of the research and development of WMDs. Similarly, members of governments have a collective moral responsibility to refrain from establishing, maintaining and/or funding WMD research and development programs. Accordingly, the relevant members of governments (as well as participating scientists and engineers etc.) have a collective moral responsibility to abandon biological weapons programs, such as that established by the Soviet Union during the communist era. Indeed, since most nation-states are signatories to the Biological Weapons Convention (BWC), this collective moral responsibility of the members of governments, and of scientists and engineers etc., is also a legal responsibility of nation-states. Further, members of the US, Russian, Chinese and other governments possessed of nuclear weapons have a collective moral responsibility to see to it that the stockpiles of these weapons are destroyed and the nuclear weapons programs abandoned by each destroying its own stockpiles of these weapons and each abandoning its own nuclear weapons programs.

Notoriously, nuclear weapons programs, in particular, give rise to collective action problems (see Chap. 6, Sect. 6.3). Perhaps one nation-state should abandon its nuclear weapons program only if other (enemy) nation-states do so, given the threat posed if the first nation-state abandons its program and its enemy nation-states do not. The mutual knowledge condition constitutive of collective ignorance is relevant here. For if each nation-state is to abandon its own program, it is crucial that the abandonment of WMD programs is verifiable; each needs to be assured of compliance by the others, if it is to comply itself. Hence the need for mutual knowledge of compliance, and hence for verification. The requirement for stringent verification procedures is part of the Chemical Weapons Convention but not, for example, the BWC. This is generally regarded as a weakness of the BWC.

I discuss such collective action problems in general terms in Chap. 4 and as they apply to particular sciences and industries (chemical, nuclear, cyber an biological) in Chaps. 5, 6, 7 and 8 respectively. Here I simply note that while such collective action problems present a challenge, they do not necessarily remove the underlying collective moral responsibility. What of dual use science and technology? Is there any collective moral responsibility to bring about collective ignorance in respect of dual use technology, specifically, dual use technology implicated in the research and development of WMD's?

As we saw above, dual use science and technology arises in the context of research in the sciences as a consequence of one and the same discrete piece, or ongoing program of scientific research, intentionally undertaken for good ends having the potential to be intentionally used for great evil. So there is an original researcher who creates new knowledge or designs new technology for good use (by the primary user), e.g. discovers how to aerosolize chemicals for use in crop dusting. But there is also a secondary user who uses the knowledge or technology for some evil purpose, e.g. uses the newly discovered process of aerosolization to weaponise chemicals.

As we saw in Chap. 2, accidents involving science and technology, even accidents on a very large scale, are not necessarily dual use in our sense since there may be no secondary evil user or, for that matter, anyone guilty of culpable negligence. Nor, as we have seen, are weapons designed as weapons, e.g. guns, instances of dual use science and/or technology.

One paradigmatic case of dual use research was the biological research done on a deadly flu virus, H5N1, which causes bird flu. In such dual use cases, the researchers—if they go ahead with the research—will have foreseeably provided the means for the evil actions of others and, thereby, arguably infringed a moral principle (albeit their infringement might in some cases be morally justified). The principle in question is the No Means to Harm (NMH) principle (elaborated in Chap. 2).[28] To reiterate (in simple terms): the NMH principle is the principle that one should not avoidably and foreseeably (whether intentionally or unintentionally) provide others with the means to intentionally do great harm.

The dual-use dilemma is a dilemma for researchers, governments, the community at large, and for the private and public institutions, including universities and commercial firms, that fund or otherwise enable research to be undertaken. Moreover, in an increasingly interdependent set of nation-states—the so-called, global community—the dual-use dilemma has become a dilemma for international bodies such as the United Nations. Accordingly, it is a matter of collective moral responsibility, and at a number of levels.

As we saw above, arguably in the case the research and development of WMDs, scientists and engineers have a collective moral responsibility (as far as is possible) to maintain or bring about collective ignorance (collective public ignorance and, with respect to certain expert groups, collective expert ignorance). Here it is important to understand how these notions of collective responsibility and collective ignorance are to be understood, and in this chapter analysis of collective ignorance has been provided—see Chap. 4 for the detailed analysis of collective responsibility. Moreover, given that research and development in dual use science and technology is likely to produce great benefits and, at least in many cases, unlikely to cause great harm,

---

[28]This principle, or similar ones, are familiar in a variety of ethical contexts. See, for example, Scanlon (1977).

there is obviously not the same or a similar collective responsibility to maintain or bring about collective ignorance as there is in the case of R&D in WMDs. On the other hand, given that *some* research and development in dual use science and technology is unlikely to produce great benefits and is likely to produce great harm, there does appear to be a collective moral responsibility to maintain or bring about some forms and degrees of collective ignorance in some dual use cases. That is, a more nuanced approach is called for. For example, arguably there was a collective moral responsibility not to undertake the above-mentioned ferret flu experiments (assuming the risk of harm was disproportionately great relative to the projected benefits) or, at least, not to publish the results in a form that would enable the experiments to be replicated (assuming publication was not necessary for the projected benefits to be forthcoming). If so, then there is a moral obligation to place restrictions on one of our identified species of collective knowledge, namely, expert knowledge, and this obligation is in turn derived from a moral obligation to maintain a specific form and degree of collective ignorance, namely, collective ignorance in sense (2) and/or sense (3), depending on whether the relevant web of knowledge is capable of being possessed by a single individual (see below).

A strong moral claim in respect of collective ignorance of harmful technology is that with respect to some WMDs (e.g. nuclear WMDs) there is a collective moral responsibility among all relevant scientists, engineers and members of other groups (e.g. government officials) to bring it about—presumably, principally via their various institutional roles—that there is mutual knowledge among them and others (including members of the public) that they and others (e.g. future scientists) do not *jointly* have the relevant web of knowledge (and cannot readily come to jointly have it). A web of knowledge was defined as an inferentially integrated cluster of molecules of propositional knowledge, acquaintance-knowledge and practical knowledge. Moreover, with respect to some other WMDs (e.g. biological WMDs) there is an analogous (putative) collective moral responsibility with respect to ensuring collective ignorance. However, in the latter case there may be the additional (difficult to realise) requirement that there be mutual knowledge that *no person individually* has the relevant web of knowledge (and cannot readily come to have it). This additional requirement is evidently superfluous in the case of nuclear WMDs because, unlike in the case of biological WMDs, it is not possible for a single individual to possess the web of knowledge in question. We pursue these matters further in subsequent chapters.

3.4 Conclusion

In this chapter I have identified three salient types [(1), (2) and (3)] and five salient species [(1a), (1b), (2), (3a) and (3b)] of collective knowledge:

(1) collective propositional-knowledge (mutual knowledge);

    (1a) public (propositional) knowledge;
    (1b) expert (propositional) knowledge;

(2) collective acquaintance-knowledge (mutual awareness);
(3) collective practical-knowledge (joint knowledge-how);

    (3a) public (practical) knowledge;
    (3b) expert (practical) knowledge.

In addition, I have defined the salient notion of collective ignorance in relation to dual use issues as follows:

Collective Public Ignorance: Members of some group, G, comprised of experts and non-experts, have collective *public* ignorance of web of knowledge, W, if and only if: there is mutual knowledge among members of G as a whole that they do not jointly have W—even if members of the sub-group of experts jointly have W—and that they cannot readily come to jointly have W by accessing available knowledge storage centres or knowledgeable persons, such as members of their expert sub-group.
Collective Expert Ignorance: Members of some group, G, comprised of experts and non-experts, have collective *expert* ignorance of web of *expert* knowledge, W, if and only if there is mutual knowledge among members of G (or, at least, among the members of the expert sub-group) that neither the members of G as a whole, nor even the members of the expert sub-group, jointly have W and that neither the members of G as a whole, nor even the members of the expert sub-group, can readily come to jointly have W by accessing available knowledge storage centres or knowledgeable persons from outside G.

Here a web of knowledge is an inferentially integrated cluster of molecules of knowledge and a molecule of knowledge is a composite of propositional, acquaintance and practical knowledge (typically) in the mind of some individual person.

Finally, I have argued that relevant scientists and technologists have a collective moral responsibility (acting jointly with others, e.g. members of governments) to maintain or bring about (i) collective *public* ignorance with respect to how to make WMDs, and also to bring it about that: (ii) no person (whether expert or not) individually knows how to make a WMD, and; (iii) members of malevolent expert groups do not jointly know how to make a WMD. Here, as elsewhere, these moral responsibilities exist only if it is possible to discharge them. In the case of some kinds of WMD this may no longer be the case.

References

Goldman, Alvin. 1999. *Knowledge in a Social World*. New York: Oxford University Press.
Hetherington, Stephen. 2011. *How to Know: A Practicalist Conception*. Oxford: Wiley-Blackwell.
Kusch, Martin. 2002. *Knowledge by Agreement*. Oxford: Oxford University Press.
Lehrer, Keith. 1990. *Theory of Knowledge*. London: Routledge.
Miller, Seumas. 1992. Joint Action. *Philosophical Papers* xxi (3): 275–299.
Miller, Seumas. 2001. *Social Action: A Teleological Account*, Chapter 8. New York: Cambridge University Press.

Miller, Seumas. 2006. Collective Moral Responsibility: An Individualist Account. In *Midwest Studies in Philosophy*, vol. XXX, ed. Peter A. French, 176–193.

Miller, Seumas. 2010. *The Moral Foundations of Social Institutions: A Philosophical Study*, Chapter 4. New York: Cambridge University Press.

Miller, Seumas. 2015. Assertions, Joint Epistemic Actions and Social Practices. In *Synthese*, published on-line April 23.

Miller, Seumas. 2017. Ignorance, Technology and Collective Responsibility. In *Perspectives on Ignorance from Moral and Social Philosophy*, ed. Rik Peels, 217–237. Oxford: Routledge.

Miller, Seumas, and Ian Gordon. 2014. *Investigative Ethics: Ethics for Police Detectives and Criminal Investigators*, Chapter 2. Oxford: Blackwell Publishing.

Peels, Rik. 2010. What is Ignorance? *Philosophia* 38 (1): 57–67.

Polanyi, Michael. 1967. *The Tacit Dimension*. New York: Anchor Books.

Popper, Karl. 1972. *Objective Knowledge: An Evolutionary Approach*, Chapter 4. Oxford: Oxford University Press.

Russell, Bertrand. 1910. Knowledge by Acquaintance and Knowledge by Description. *Proceedings of the Aristotelian Society* 11: 108–128.

Scanlon, Thomas. 1977. A Theory of Freedom of Expression. In *The Philosophy of Law*, ed. Ronald M. Dworkin. Oxford: University Press.

Schmitt, Frederick (ed.). 1994. *Socializing Epistemology: The Social Dimensions of Knowledge*. Lanham: Rowman and Littlefield.

Smith, N.V. (ed.). 1982. *Mutual Knowledge*. London: Academic Press.

Unger, Peter. 1974. *Ignorance: The Cases for Scepticism*. Oxford: Oxford University Press.

Chapter 4
Collective Responsibility

Abstract Scientific freedom is rightly extolled as an important moral and intellectual value. However, as is often noted, with freedom comes responsibility; scientific freedom is no different in this respect. However, science is essentially a cooperative enterprise that typically takes place in institutional settings and is shaped by institutional purposes. Therefore, scientific freedom, properly understood, is in large part an expression of the intellectual freedom of scientists engaged in cooperative epistemic activity in organisations, such as universities and firms. Accordingly, the responsibilities of scientists and technologists are a species of *collective responsibility*. In this chapter I argue that collective responsibility is essentially joint responsibility and, therefore, a species of relational individual human responsibility. I distinguish institutional responsibility from moral responsibility, and both from mere causal responsibility. In doing so I also rely on the analysis of the organized, indeed organisational, action of scientists in terms of my notion of a layered structure of joint epistemic actions. This analytical notion allows me to ascribe collective moral responsibility to scientists, at least in principle, both for the bad, as well as the good, outcomes of their research. It paves the way for scientists and technologists to be ascribed moral responsibility (jointly with legislators, regulators etc.) for devising training programs, regulations and so on to deal with dual use issues. I also consider various collective actions problems that exacerbate dual use problems.

Scientific freedom is rightly extolled as an important moral and intellectual value. However, as is often noted, with freedom comes responsibility; scientific freedom is no different in this respect. However, as we saw in Chap. 3, science is essentially a cooperative enterprise; indeed, one that typically takes place in institutional settings and is shaped by institutional purposes. Therefore, scientific freedom, properly understood, is in large part an expression of the intellectual freedom of scientists engaged in cooperative epistemic activity in organisations, such as universities and firms. Accordingly, the responsibilities of scientists and technologists are a species of *collective responsibility*. But how is the somewhat opaque notion of collective responsibility to be understood? In this chapter I argue that collective responsibility is essentially joint responsibility and, therefore, a species of relational individual

© The Author(s) 2018

S. Miller, *Dual Use Science and Technology, Ethics and Weapons of Mass Destruction*, SpringerBriefs in Ethics, https://doi.org/10.1007/978-3-319-92606-3_4

human responsibility. In doing so I distinguish institutional responsibility from moral responsibility, and both from mere causal responsibility. In doing so I also rely on the analysis of the organized, indeed organisational, action of scientists in terms of my notion of a layered structure of joint epistemic actions. This analytical notion allows me to ascribe collective moral responsibility to scientists, at least in principle, both for the bad, as well as the good, outcomes of their research. So it paves the way for scientists and technologists to be ascribed moral responsibility (jointly with legislators, regulators etc.) for devising training programs, regulations and so on to deal with dual use issues. In the final section of this chapter I consider various collective actions problems that exacerbate dual use problems. In later chapters I argue that the generic solution to these collective action problems involves designing and implementing institutional arrangements that one way or another embed collective responsibilities in enforceable cooperative schemes.

4.1 Scientific Freedom, Joint Action and Organisational Action

According to scientist-cum-philosopher Michael Polanyi:

> The existing practice of scientific life embodies the claim that freedom is an efficient form of organisation. The opportunity granted to mature scientists to choose and pursue their own problems is supposed to result in the best utilization of the joint efforts of all scientists in a common task. In other words: if the scientists of the world are viewed as a team setting out to explore the existing openings for discovery, it is assumed that their efforts will be efficiently coordinated if only each is left to follow his own inclinations. It is claimed in fact that there is no other efficient way of organizing the team, and that any attempts to coordinate their efforts by directives of a superior authority would inevitably destroy the effectiveness of their cooperation.[1]

Polanyi's view is each scientist acts freely but does so:

(1) on the basis of the work of past scientists;
(2) with constant reference and adjustment to the work of other contemporary scientists; and
(3) in the overall service of a collective end of comprehensive knowledge (in the sense of understanding) of the scientific phenomena in question.

So his conception is one of individual scientific freedom in the overall context of intellectual interdependence in a joint knowledge-aiming or *epistemic* project—a project of joint epistemic action.[2] Hence the knowledge aimed at is collective knowledge. Moreover, as we saw in Chap. 3, Sect. 3.1, the collective knowledge in question is an integrated mix of knowledge by acquaintance, practical knowledge and propositional knowledge and, specifically, of (at least) collective

---

[1] Polanyi (1951), 34.
[2] Miller (2015a, 280–302) and Miller (2016).

expert propositional-knowledge and collective expert practical-knowledge, i.e. it is collective expert knowledge.

So joint epistemic action is knowledge acquisition involving multiple epistemic agents seeking to realize a collective epistemic end. For example, the members of a team of scientists seeking knowledge of the cure for cancer are engaged in joint epistemic action.[3] In cases of joint epistemic action there is mutual true belief among the epistemic agents that each has the same collective epistemic end, e.g. to discover the cure for cancer. Moreover, there is typically a division of epistemic labour. Thus in scientific cases some scientists are engaged in devising experiments, others replicating experiments, and so on. So, as is the case with joint action more generally, joint epistemic action involves interdependence of individual action—albeit interdependence of individual epistemic action—and the pursuit of collective ends.

A collective epistemic end can be both a collective intrinsic good—and thus an end-in-itself—and also the means to further ends. Knowledge of the cure for cancer—indeed, collective expert knowledge of the cure for cancer—is a case in point. Such collective knowledge consists of propositional and practical knowledge; knowledge of the cure for cancer and knowledge how to produce it. Moreover, the knowledge in question is collective knowledge, specifically, collective expert knowledge (see Chap. 3, Sects. 3.1 and 3.2). However, acquisition of this collective knowledge serves a further (collective) end, namely, the production of the actual physical cure, such as a drug. And this end has in turn a still further end, namely, to save lives. If collective expert knowledge of the cure for cancer is a collective end in itself then it is not simply a means to individual ends, viz. each having as an end that he or she alone knows the cure for cancer. Rather it is mutually truly believed that collective expert knowledge of the cure for cancer is a collective end-in-itself. Moreover, it is a collective good which is a means to a further good, namely, the saving of lives. Here the saving of lives is an intrinsic good but so also, it would be argued by many scientists, is collective knowledge of the cure for cancer; scientific knowledge being an intrinsic, and not merely instrumental, good.

I have been stressing the cooperative and collective character of scientific knowledge (and the exercise of scientific freedom). It is now time to emphasise the institutional character of science. As we have seen, scientific research consists of joint epistemic action. The joint epistemic action in question typically take place in institutional settings, such as universities, commercial firms and government research facilities. Indeed, this joint epistemic action is institutionally embedded and, as such, is itself a species of organisational action. Thus, the research to be undertaken is in large part institutionally determined (including by means of economic incentives[4]), the scientists and technologists (knowledge workers, so to speak) are organized hierarchically and according to principles of the division of labour, the epistemic fruits of the research are typically commercialised or otherwise utilized in accordance with institutional directives, (e.g. from government), and so on.

---

[3] See Miller (2010, Chap. 11).
[4] Resnik (2007).

Organisational action typically consists of, what elsewhere I have termed, a *layered structure of joint actions*.[5] Importantly for our purposes here there are layered structures of joint *epistemic* action. Consider a crime squad, comprised of detectives, forensic scientists etc., attempting to solve a crime.[6] At level one, a victim, A, communicates the occurrence of the crime (say, an assault) and description of the offender to a police officer, B. But A asserting that p to B is a joint epistemic action; it is a cooperative action governed by social norms and conventions, such as the social norm that the speaker A tells the truth and the hearer trusts the speaker to tell the truth.[7] Also at level one, a couple of detectives interview the suspect to determine motive and opportunity; the detectives are cooperating with one another in the performance of a joint epistemic action the collective end of which is to discover, for instance, motive and opportunity. Finally, at level one, a team of forensic scientists analyze the available physical evidence e.g. the DNA of the blood samples of the offender found on the victim are matched to the suspect's DNA; the forensic scientists are engaged in joint epistemic action to determine whether there is or is not a DNA match. These three level one joint epistemic actions are constitutive of a level two joint epistemic action, namely, the level two joint epistemic action directed towards the collective end of determining who committed the crime. Accordingly, when each of the level one joint epistemic actions is successfully performed then the level two joint epistemic action is successfully performed, i.e. the crime squad—comprised of police officers, detectives and forensic scientists—solves the crime.

Now consider an example of a large scientific project conducted by a number of cooperating organisations (principally 20 universities and research centres) and hundreds of scientists over many years (roughly from 1988–2001), namely, the Human Genome Project (HGP). HGP was the international, collaborative research program whose collective epistemic end was the complete mapping and understanding of all the genes of human beings, i.e. the human genome. According to the National Human Genome Research Institute, "The HGP has revealed that there are probably about 20,500 human genes. The completed human sequence can now identify their locations. This ultimate product of the HGP has given the world a resource of detailed information about the structure, organisation and function of the complete set of human genes. This information can be thought of as the basic set of inheritable "instructions" for the development and function of a human being."[8] Accordingly, the realised collective end of the project was collective expert knowledge of the human genome, i.e. a web of knowledge (see Chap. 3, Sects. 3. 1 and 3.2). This web of knowledge consists of fragments of knowledge and these fragments were the epistemic contributions of multiple researchers working in multiple different organisations world-wide. So HGP involved realizing multiple, nested, collective epistemic ends (fragments of knowledge) in the service of the larger collective epistemic end of mapping and understanding the human genome

---

[5]Miller (2001).

[6]Miller and Gordon (2014, Chap. 2).

[7]Miller (2010, Chap. 11), Miller (2015b), Miller (1992b, 435–445) and Miller (1997, 211–229).

[8]See the homepage of the National Human Genome Research Institute at www.genome.gov.

(web of knowledge), and multiple layered structures of joint epistemic action undertaken to realize this larger collective epistemic end.

To sum up: the scientific enterprise is a species of organisational action involving layered structures of joint epistemic action. Moreover, the organisations in question are, for the most part, hierarchical institutions comprised of task-defined roles standing in authority relations to one another, designed in accordance with principles of division of labour and governed by a complex network of conventions, social norms, regulations and laws. Consider a science department in a university or the forensic laboratory in a police organisation: both comprise heads of department, scientists, laboratory assistants, and so on, and the work of both is governed by scientific norms of observation, replication of experiments etc.

Let me conclude this section with some observations about institutions including those undertaking scientific research. Institutions have de facto purposes/strategic directions, i.e. collective ends, such as to maximize shareholder profit (corporations), find a cure for cancer (university research team), design an atomic bomb (military organisation). Institutions also have specific structures (hierarchical, collegial etc.) and they have specific cultures (e.g. a competitive, status-driven ethos).[9] In this connection, consider scientific activity, e.g. biological research, undertaken in three different institutional settings—that of the university, the commercial firm and the military bio-defence organisation. Some of the principal purposes/strategic directions (collective ends) of commercial firms, e.g. to maximize shareholder profits, are different from, and possibly inconsistent with, those of universities, e.g. scientific knowledge for its own sake, and quite different again from those of military research establishments, e.g. to save the lives of the military personnel of the nation-state in question in times of war.[10] Again, the hierarchical structures within a military research establishment are quite different from the more collegial structures prevailing in universities; and the structure of commercial firms is quite different again. The general point to be made here is that scientific activity is not only a form of complex joint activity (a layered structure of joint epistemic action), it is activity that is inevitably shaped by the institutional setting in which it is conducted, i.e. by the specific collective ends, structure and culture of the institution in which it is embedded.

Here we need to stress the distinction between the de facto (what in fact is the case) and the normative (what ought to be). The de facto institutional collective end, structure, and/or culture may not be what it ought to be (see Sect. 4.2). We can also distinguish the normative account of science as a joint intellectual activity, e.g. aimed at knowledge for its own sake, from science as means to broader social ends, e.g. vaccines to save lives. Moreover, we can distinguish both from the normative account of specific institutions in which science exists principally as a means, e.g., military

---

[9]Miller (2010).

[10]This is not to deny that these different institutions interact and influence one another so that, for instance, university-based research is not at times indirectly driven by the profit motive of a commercial firm. And, of course, universities also seek to make money directly from their scientific research, e.g. by commercialising it via intellectual property rights.

bio-defense organisation (vaccines to save the lives of the military personnel of the nation-state in question).

Importantly, in the context of discussion of dual use concerns, we can distinguish within the normative account of science (both at the level of joint intellectual activity and at the level of specific institutions) between its beneficial ends, (e.g. knowledge for its own sake and knowledge as a means to combat disease), and its accompanying side-constraints, (e.g. avoid harming humans in this process of pursuing beneficial ends). In Chap. 2, Sect. 2.3 I introduced the No Means to Harm principle (NMH), namely, the principle that one should not avoidably and foreseeably (whether intentionally or unintentionally) provide others (e.g. bioterrorists) with the means to do great harm. Clearly the means in question is essentially scientific or technological knowledge and this knowledge is the product of joint epistemic action (indeed, joint epistemic action undertaken in the context of multiple layered structures of joint epistemic action). Accordingly, harm prevention in relation to dual use concerns is, or at least morally ought to be, a joint endeavour of scientists; it is something that scientists as members of their scientific community (or communities) and specific institutions (e.g. biology departments in universities, biotech companies), morally ought to jointly address including (presumably) by way of education and regulation of their potentially harmful joint epistemic action. Accordingly, scientists have responsibilities in relation to dual use concerns. Moreover, given the essentially cooperative and, as it now turns out, institutional character of the scientific enterprise, these responsibilities are, on the one hand, collective responsibilities and, on the other, institutional responsibilities. This raises the question of the relationship between institutional and moral responsibility.

4.2   Institutional and Moral Responsibility

Institutional responsibility contrasts with both natural and moral responsibility.[11] Natural responsibility does not depend on one's institutional role and is not necessarily moral in character. If Jones intentionally crosses the road to get to the other side then he is naturally responsible for this fact; but his action is not necessarily moral in character (let alone dependent on, or directly relevant to, any institutional role he might have). Now consider moral responsibility. Jones might be morally responsible for failing to assist a frail old woman to cross the road. Notice that he might be morally responsible for this omission without being institutionally responsible for it. Equally, Smith might be institutionally responsible for seeing to it that her desk is tidy but we might baulk at regarding this as a moral responsibility. Moreover, responsibility can be used in a backward or a forwarding looking sense. An example of the former sense is: 'Jones is responsible for the car crash since he failed to stop at the red traffic light'. An example of the latter sense is 'The mechanic is responsible for seeing to it that the brakes in my car are fixed'. Notice that in the case of back-

---

[11] Miller (2017, 338–348).

ward looking responsibility at least, we must distinguish mere causal from moral responsibility (and, for that matter, from institutional responsibility). This if Jones caused the car crash but was unconscious at the time due to a sudden heart attack then he was presumably not morally responsible for it, notwithstanding his causal responsibility.

Responsibility needs to be distinguished from blameworthiness/praiseworthiness, on the one hand, and accountability, on the other. If a Ph.D. student performs his allotted task of conducting a routine experiment to an acceptable standard, but not to a high standard, then he is responsible for having conducted the experiment; but he is presumably neither praiseworthy nor blameworthy. Evidently, therefore, praiseworthiness and blameworthiness presuppose responsibility, but should not be equated with it. Again, responsibility should not be confused with accountability. The Ph.D. student is responsible, let us assume, for conducting the routine experiment, but he is accountable for his performance as a Ph.D. student to (say) his supervisor. That is, the supervisor might be tasked with monitoring and assessing the student's performance and, if necessary, intervening in the case of poor performance by retraining, disciplining or perhaps even recommending to the university that his candidature be terminated.

The notion of institutional responsibility presupposes some notion of an institution. In this work the focus is only with institutions that are also organisations and/or systems of organisations and, as we have seen, institutions have purposes (collective ends) and role-based structures. The third main dimension of institutions is culture; the 'spirit' or informal set of attitudes that pervades an organisation and which might reinforce or negate the more formal requirements of the organisation.

One normative theory of social institutions is based on an individualist theory of joint action.[12] Put simply, on this account institutions are organisations or systems of organisations that provide collective goods by means of joint activity. So on this account institutional purposes are collective ends that are collective goods. The collective goods in question include the fulfilment of aggregated moral rights, such as needs based rights for security (police organisations). In the case of universities and other research institutions, the collective goods in question are epistemic goods, e.g. knowledge of atoms, numbers or historical periods.

In the light of the above, we can distinguish three possible ways of understanding institutional responsibility. Firstly, there is the responsibility *to institutions*. This is the responsibility (possibly moral responsibility) that a single individual or, more likely, members of a group might have to establish, maintain or redesign an institution. Here the property 'institutional' does not qualify the notion of responsibility; rather it is part of the content of the responsibility. This sense of institutional responsibility is not our concern in this chapter. Secondly, there is responsibility *of institutions*. This is the notional possibility that institutional responsibility might attach to collective entities (specifically, institutions) per se. This possibility could only obtain if institutions (and like collective entities) were *minded* agents: agents possessed of mental states, such as desires, intentions, and beliefs. For only minded agents perform actions in the

---

[12]Miller (1992a), Miller (2001, Chap. 2) and Miller (2010, Chap. 2).

appropriate sense of action, and only minded agents can sensibly be held responsible for their actions. However, the idea that institutions per se, as opposed to their human members (institutional role occupants), have minds is problematic or, at the very least, controversial. At any rate, in this work I set aside any further consideration of this way of understanding institutional responsibility. Thirdly, there is the responsibility *of institutional role occupants*. This is the institutional responsibility of the human beings who occupy institutional roles. This is the sense of institutional responsibility of interest to us.

Evidently individual role occupants are *individually* institutionally responsible for at least some of their actions and omissions. For instance, the above-mentioned Ph.D. student was individually institutionally responsible for conducting the experiment (responsibility in the forward looking sense). Moreover, if the student fails to conduct the experiment to the required standard then he is individually responsible for not having conducted it properly (backward looking sense) and this failure attaches to him qua institutional role occupant (Ph.D. student).

Again, an institutional role occupant in a position of authority over another (e.g. the supervisor in relation to the Ph.D. student) might have an *individual* institutional responsibility (forward looking sense) to see to it that her subordinate performs the tasks definitive of the subordinate's role. Moreover, if the subordinate (the student) consistently fails to perform the tasks in question, and his superior (the supervisor) fails to intervene, then the supervisor is individually responsible (backward looking sense) for failing to see to it that the student does his work and this failure attaches to the supervisor qua institutional role occupant.

On the other hand, a number of institutional role occupants might be *collectively* institutionally responsible for some outcome. The paradigmatic cases here are ones of joint action, including joint epistemic action; actions involving cooperation between institutional actors to achieve some, possibly epistemic, outcome. How are we, then, to understand the notion of collective responsibility?

4.3 Collective Responsibility

Collective responsibility of the kind in question here is the responsibility that attaches to the participants of a joint action for the performance of joint action and, in particular, for the realisation of the collective end of the joint action (including joint omission understood as involving intentions, or otherwise aiming, to refrain from action). There are different accounts of collective responsibility, some of which pertain to the responsibility of groups and organisations per se for their group or 'corporate' (so to speak) actions. Here our concern is only with collective responsibility for joint actions of human beings in their capacity as institutional role occupants. As already mentioned above, one such salient account conceptualises collective moral responsibility for joint action as *joint responsibility*.[13]

---

[13]Miller (2006).

On this view of collective responsibility as joint responsibility, collective responsibility is ascribed to individual human beings only, albeit jointly.[14] Moreover, institutional actors can be ascribed collective institutional responsibility when they act jointly in accordance with their institutional roles. Consider a fire-fighting team at a toxic waste storage site. Each member of the group is individually institutionally responsible for their contributory action and also for the aimed at outcome (the collective end) of the set of actions. However, each fire officer is individually responsible for that outcome, *jointly with the others*; so the conception is relational in character. Thus in this fire-fighting example, each member of the teams is institutionally responsible jointly with the others for extinguishing the fire because each performed his or her contributory action in the service of that collective end (putting out the fire). So the members of the fire-fighting team are collectively institutionally responsible for extinguishing the fire (initially in the forward-looking sense of responsibility and after the fire has been extinguished in the backward-looking sense).

What of the collective responsibility of institutional actors engaged in epistemic enterprises? Recall the Human Genome Project. As argued above, the scientists engaged in the HGP were participating in a layered structure of joint epistemic action the collective end of which was a web of knowledge: collective expert knowledge of the human genome. Accordingly, each participating scientist in a given team that contributed a fragment of this web of knowledge can be held individually responsible for whatever epistemic contribution he or she individually made, i.e. for his or her 'piece' of the fragment. In addition, each participating scientist in a given team that contributed a fragment of this web of knowledge can be held individually responsible, *jointly with the others*, for that fragment of knowledge. Further, each participating scientist in a given team that contributed a fragment of this web of knowledge also contributed to the web of knowledge of which that fragment of knowledge was a fragment. Accordingly, *each contributing scientist can be held individually responsible, jointly with the others, for the web of knowledge*, i.e. for the map of the human genome. Naturally, this latter responsibility is not full, but only partial, individual responsibility; each had a (typically small) share of the overall responsibility. No single scientist can take *full* responsibility for mapping the human genome; but nor is it the case that no single individual scientist can take *any* responsibility for mapping the human genome.

Thus far we have distinguished between natural, institutional and moral responsibility and, in respect of responsibility, between individual and collective responsibility. I note that the notions of natural, institutional and moral responsibility are not mutually exclusive, but how are they related?

To recap. An agent, A, has *natural* responsibility for some action, x, if A intentionally did x for a reason and x was under A's control. Bench scientists engaging in routine scientific research, e.g. replication of experiments, have natural responsibility for their actions. Moreover, such actions might not have any obvious moral

---

[14] Accordingly, there is no need to hold that collective responsibility attaches to collective entities per se, as collectivist theorists such as Margaret Gilbert and (in a somewhat different vein) Philip Pettit have done. For criticisms of these collectivist accounts see Miller and Makela (2005, 634–651).

implications. Agent A has *institutional* responsibility for action x if A has an institutional role that has as one of its tasks to x. Thus, for example, laboratory assistant, A, has the institutional responsibility to clean the test tubes; moreover, A has this responsibility even if A does not in fact do this. What of moral responsibility?

Roughly speaking, agents have moral responsibility for natural or institutional actions if those actions have moral significance. So if A is naturally or institutionally responsible for x (or for some foreseeable outcome of x, O) and x (or O) is morally significant then—other things being equal—A is morally responsible for x (or O) and—other things being equal—can be praised/blamed for x (or O).

Note that other things might not be equal if, for example, A is a psychopath (and, therefore, incapable of acting in a morally responsible fashion) or if A does something wrong but has a good excuse (and, therefore, ought not to be blamed). Note also that if O involves some intervening agent, B, who directly causes O then A may have diminished moral responsibility for O.

Let us now consider collective moral responsibility. In essence, the account of collective moral responsibility mirrors that of individual moral responsibility, the key difference being that the actions in question are joint actions, including joint epistemic actions. It also needs to be borne in mind that the joint epistemic actions in question might comprise layered structures of joint action—as in the case of the HGP described above—in which case, as we saw, the collective responsibility in question is the joint responsibility of all (or, at least, most) of the participants in the larger structure of joint epistemic actions. I use the term "joint activity" in the definition below to refer to layered structures of joint action as well as single joint actions.

Accordingly, if agents, A, B, C etc. are naturally or institutionally responsible for a joint (including epistemic) activity x (and/or some foreseeable outcome of x, O) and x (and/or O) is morally significant then—other things being equal—A, B, C etc. are collectively (i.e. jointly) morally responsible for x (and/or O) and—other things being equal—can be praised or blamed for x (and/or O).

The 'other things being equal' clauses function here as they did in the above account of individual moral responsibility. Moreover, as was seen to be the case with individual moral responsibility, if there are additional intervening (individual or joint) actions then those jointly responsible for the joint action in question, and its outcome, may have diminished moral responsibility. Scientists who engage in dual use research which is subsequently used in the construction of WMD's may well have diminished responsibility for the harm caused by those WMD's. However, diminished responsibility is not necessarily equivalent to no responsibility. Further points to be made here are as follows.

First, each agent may have full or partial moral responsibility for x jointly with others for the joint action x and/or its outcome. If, for example, five men each stab a sixth man once killing him, each is held *fully* morally (and legally) responsible for the death even though no single act of stabbing was either necessary or sufficient for the death. In some cases each agent might have full moral responsibility (jointly with others) for some outcome O—notwithstanding the fact that each only made a very small causal contribution to the outcome—in large part because each is held to have prior full institutional (including legal) responsibility (jointly with others) for O.

On the other hand, each agent might have partial and minimum moral responsibility jointly with others if each only makes a very small and incremental contribution as a member of a very large set of agents performing their actions over a long period of time, e.g. the scientists who worked on the HGP. Moreover, in hierarchical organisations, individual scientists operating under the authority of other scientists or managers might only have diminished (partial) responsibility relative to those in authority.

Second, we need to distinguish cases in which agents have collective moral responsibility for some joint action or its outcome from cases in which agents only have collective moral responsibility for failing to take adequate preventative measures against O taking place. Many untoward dual use cases are of the latter kind.

Agents may not have any collective (or individual) moral responsibility with respect to some foreseeable morally significant outcome, O, if O has a low probability, takes place in the distant future and involves a large number of intervening agents. That said, the analytical notion of a layered structure of joint action, taken in conjunction with the notion of collective responsibility as joint responsibility, allows me to ascribe collective moral responsibility to scientists operating in complex organisations, at least in principle; and it does so for the bad as well as the good outcomes of their research. In doing so it paves the way for scientists and technologists to take moral responsibility (jointly with legislators etc.) for devising training programs, regulations and so on to deal with dual use iss ues.

The collective moral responsibilities of scientists are multiple. Scientists have a collective institutional (professional) and moral responsibility as scientists to acquire knowledge for its own sake. Scientists functioning in universities also have a collective institutional and moral responsibility to acquire knowledge for the good of humanity, e.g. vaccines for poverty-related diseases. Scientists functioning in commercial firms might have a collective institutional and (contractually based) moral responsibility to acquire (say) knowledge of vaccines for rich people's diseases—since that is a commercial imperative of their employer and they are being paid to do just that. Scientists functioning in bio-defense organisations have a collective institutional (and moral?) responsibility to acquire knowledge of vaccine resistant pathogens if this is a national security imperative of their employer, viz. the government. As human beings scientists have a collective moral responsibility not to provide the means for others to intentionally do great harm, e.g. the means to allow others to drop atomic bombs on Hiroshima and Nagasaki[15] or engage in bio-warfare.

Moreover, these various collective institutional and moral responsibilities may be inconsistent with one another, notably the collective moral responsibilities scientists have as human beings and the institutional responsibilities that they might have as members of military research organisations.

---

[15]Some have argued (controversially) that dropping atomic bombs on Hiroshima and Nagasaki by the US military, while extraordinarily harmful, was morally justified all things considered (e.g. because it reduced the loss of life overall or because it spared US military losses in particular). If this argument is sound then the collective responsibility not to provide others to do great harm might have been overridden *in this instance*.

4.4   Collective Action Problems

Thus far we have characterized the scientific enterprise as essentially a joint epistemic enterprise: the emphasis has been on intellectual cooperation to achieve common scientific (epistemic) goals in an institutional context defined in part by those goals.[16] However this picture, while acceptable as far as it goes, is an oversimplification. Specifically, it obscures the competitive dimension of scientific activity and, in particular, it masks various collective action problems arising from such competition. This is important for our purposes here, not the least because it casts dual use problems in a somewhat different light and, in some cases, may well exacerbate them.

On the purist (as we might call it) model of scientific activity as joint epistemic action performed under conditions of scientific freedom, the dual use problem arises only because scientific research undertaken for the benefit of humankind can be misused by others for harmful purposes. Accordingly, there is a need to monitor dual use research and erect safeguards against misuse by malevolent individuals and groups, e.g. 'lone wolf' malcontents, nihilistic terrorist groups, 'rogue states' and so on.

Notice, firstly, that there is here an implicit additional assumption, namely, that scientific activity will be undertaken in the first place in order to benefit humankind. This is, as we have seen in relation to WMD programs, not necessarily the case. On the other hand, WMD research is not dual use in our sense (unless it is military research undertaken for purely protective, as opposed to deterrence, purposes—yet having the potential (as is probable) to be misused for aggressive purposes).

Notice, secondly, that much scientific work, including not only in the chemical, nuclear and cyber fields, but also in the biological sciences, is not undertaken under conditions of scientific freedom, at least in any strong sense of that term. Consider research undertaken in the private sector or for various government laboratories in, for example, authoritarian states such as China. Which research is undertaken, and whether or not it is published, are not necessarily or even typically decisions made by individual scientists or, indeed, by groups of scientists. Rather these are commercial decisions made by managers or they are decisions made by government officials in the national interest (presumably). Accordingly, it is simply not true that scientific work, including scientific work in the biological sciences, let alone in the chemical, nuclear and cyber fields, is necessarily, or even typically, conducted under conditions of scientific freedom (in any strong sense).

But to return to the main point at issue, namely, competition and, relatedly, collective action problems: In the biological sciences, as elsewhere, there is competition between individual scientists, between scientific institutions and between nation-states.

It is self-evident that there is competition between, for example, biotechnology companies in the private sector. Moreover, governments compete in so far as they have an interest in promoting their own biotech industries and, more generally, in so far as they want to ensure that they do not fall behind in R&D in the various scientific

---

[16] An earlier version of the material in this section appeared in Miller (2013, 185–206).

and technological areas in question (not the least for military reasons). Further, even in the case of scientific work undertaken under conditions of scientific freedom, e.g. in universities, there are important elements of competition, e.g. between rival teams of scientists in competition for status and (relatedly) for scarce funding[17].

As suggested above, competition in these various sectors gives rise to a variety of collective action problems that have important implications for the dual use issue. First, in the private sector there are collective action problems arising from commercial competition. As already noted, many scientists work in commercial firms in which there is an imperative to maximise profit. In such a context of fierce commercial competition restrictions on dual use research may handicap an organisation. This is a collective action problem in so far as an organisation—all things considered—ought to choose *not* to perform a particular dual use experiment on the grounds that the potential harm to humankind resulting from this kind of experiment might outweigh the potential benefits to humankind. However, all moral things might not be considered or (if considered) given appropriate weight. Specifically, the firm might give excessive weight to its commercial interests, especially if it believes that some other competing firm is likely to be less scrupulous and go ahead with the experiments in question. In short, in dual use cases where discretionary judgment is called for, the judgment might be skewed by considerations of commercial self-interest in a fiercely competitive commercial environment. I note that commercial self-interest may well be dominant in such cases, notwithstanding the commitment of individual scientists to the No Means to Harm principle. For one thing, it is not necessarily a matter for the decision of the scientists—who are, after all, mere employees; and for another, their self-interest as employees might align them with the firm's commercial interest, especially given the relative ignorance of scientists of security issues.

Second, in the university sector there are collective action problems arising from competition for status. As already noted, many scientists working in the university sector are engaged in a competition for status (and for scarce funds to undertake projects by means of which they can achieve status), both for themselves as individuals and on behalf of the institutions they work for. Accordingly, there is an analogue in the university sector of the above-described collective action problem that, as we saw, arises in the private sector. In dual use cases where discretionary judgment is called for, the judgment might be skewed by considerations of individual or institutional self-interest in a competitive environment, albeit the competition in universities is primarily for status (and scarce funds to achieve status).

In later chapters, especially Chaps. 6 and 8, I argue, in effect, that we require a web of institutional arrangements to deal with these first two collective action problems (and related ones), e.g. regulations, training programs: in short, the collective (i.e. joint) moral responsibility of scientists and others needs to be embedded in institutional arrangements that are, in effect, enforceable cooperative schemes.

Thirdly, in the government sector there are collective action problems arising from competition among nation-states. As noted above, in the past and, indeed, in the present there have been a variety of arms races, e.g. the nuclear arms race, in which

---

[17]Resnik (2007).

scientists played a central role. The problem here is that national self-interest is pitted against humanity's collective interest in a context in which there is no enforceable international law; evidently nation-states cannot effectively collectively self-regulate. Hence the WMD programs of the US, Iran, North Korea and so on. However, the inability or unwillingness to collectively self-regulate exists, at least potentially, in relation to the dual use problem and does so independently of any desire on the part of nation-states to maintain WMD programs. We saw above that the self-interest of individual scientists and the institutions in which they work (e.g. commercial firms and universities) can under conditions of fierce competition lead to collective action problems in relation to dual use research. However, nation-states are themselves in competition with one another, and it is typically in the economic, military and political interests of nation-states to support their own R&D in science and technology, i.e. to support the work of their commercial firms and universities, and do so in the face of 'foreign' competition. Accordingly, we cannot necessarily look to individual governments to regulate adequately the scientific research in their own institutions, at least if "adequately" in this context refers to an all things considered morally, as well as empirically, informed decision made in the long term interests of humankind—as opposed to a decision made in the (possibly short term) national interest.

In Chap. 6 and elsewhere in this work I argue, in effect, that we require enforceable cooperative schemes to deal with this third kind of collective action problem. In these cases the cooperative schemes in question are ones to which all or most governments need to sign up to. However, the problems arise with enforcement, as we shall see in Chap. 6. At any rate, the point is that the prior collective (i.e. joint) moral responsibility of the members of national governments needs to be embedded in institutional arrangements.

A fourth collective action problem arising from competition is of a somewhat different kind; it is a species of the generic problem of free-riding. It is the possibility of the untoward consequences of scientific free-riding, so to speak. Let us assume that Polanyi's scientific freedom model, e.g. no censorship, is in fact the best model to acquire new knowledge; those operating entirely outside the model cannot compete. Accordingly, so the argument runs, the 'good guys' (e.g. the scientists making vaccines within the framework of scientific freedom) stay ahead of the 'bad guys' (e.g. the scientists weaponising pathogens outside the framework of scientific freedom); the 'bad guys' are always playing catch-up. However, contrary to this argument, it might be claimed that a well-qualified national cohort of 'bad guys' can always free ride but then get ahead of 'good guys', e.g. scientists in an authoritarian state with bio-defense projects benefit from work of those in scientific freedom model but don't share their own work. Arguably, North Korea's currently very successful (in terms of the threat it poses) nuclear weapons program is a case in point. It is now widely believed that it threatens the US mainland.

4.5   Conclusion

The exercise of individual scientific freedom takes place in an overall context of intellectual interdependence in a joint knowledge-aiming or *epistemic* project—a project of joint epistemic action aiming at collective knowledge (or, at least, collective expert knowledge). The joint epistemic projects in question not only typically take place in institutional settings, such as universities and private firms, the joint epistemic action is itself a species of organisational action, i.e. the scientists and technologists are organized hierarchically, according to principles of the divisions or labour, and so on. Organisational action typically consists of layered structures of joint actions and epistemic institutions of *layered structures of joint epistemic action*. As such, scientists and technologists engagement in R&D is a species of joint epistemic action and they can be held collectively, i.e. jointly, institutionally and morally responsible for its bad outcomes as well as its good outcomes, at least in principle. Accordingly, scientists and technologists must accept a collective responsibility (jointly with legislators etc.) to design and implement training programs, regulations and so on to deal with dual use issues.

I have defined the notion of collective moral responsibility as follows: if agents, A, B, C etc. are naturally or institutionally responsible for a joint (including epistemic) activity x (and/or some foreseeable outcome of x, O) and x (and/or O) is morally significant then—other things being equal—A, B, C etc. are collectively (i.e. jointly) morally responsible for x (and/or O) and—other things being equal—can be praised or blamed for x (and/or O). The joint epistemic action of scientists and technologists in the institutional contexts in question (universities, firms, government research agencies) gives rise to various collective action problems that may exacerbate dual use problems, e.g. the problem of free-riding malevolent secondary users.

References

Miller, Seumas. 1992a. Joint Action. In *Philosophical Papers*, vol. xxi no. 3, 275–299.

Miller, Seumas. 1992b. On Conventions. *Australasian Journal of Philosophy* 70 (4): 435–445.

Miller, Seumas. 1997. Social Norms. In *Contemporary Action Theory (Volume 2: Social Action)*, ed. G. Holmstrom-Hintikka, and R. Tuomela, 211–229. Dordrecht: Kluwer-Synthese Library Series.

Miller, Seumas. 2001. *Social Action: A Teleological Account*, Chapter 2 and Chapter 5. New York: Cambridge University Press.

Miller, Seumas. 2006. Collective Moral Responsibility: An Individualist Account. In *Midwest Studies in Philosophy*, ed. Peter A. French. Vol. XXX, 76–193.

Miller, Seumas. 2010. *The Moral Foundations of Social Institutions: A Philosophical Study*. New York: Cambridge University Press (Introduction, Chapter 2, Chapter 4, Chapter 11).

Miller, Seumas. 2013. Collective Responsibility, Epistemic Action and the Dual Use Problem in Science and Technology. In *On the Dual Uses of Science and Ethics: Principles, Practices and Prospects*, ed. Brian Rappert, and Michael Selgelid, 185–206. Canberra: ANU Press.

Miller, Seumas. 2015a. Joint Epistemic Action and Collective Moral Responsibility. *Social Epistemology* 29 (3): 280–302.

Miller, Seumas. 2015b. Assertions, Joint Epistemic Actions and Social Practices. In *Synthese*, published on-line 23rd April 2015.

Miller, Seumas. 2016. Joint Epistemic Action: Some Applications. *Journal of Applied Philosophy*, published on-line February 25.

Miller, Seumas. 2017. Institutional Responsibility. In *Routledge Handbook of Collective Intentionality*, eds. Kirk Ludwig and Marija Jankovic, 338–348. Oxford: Routledge.

Miller, Seumas and Ian Gordon. 2014. *Investigative Ethics: Ethics for Police Detectives and Criminal Investigators,* Chapter 2. Oxford: Blackwell Publishing.

Miller, Seumas, and Pekka Makela. 2005. The Collectivist Approach to Collective Moral Responsibility. *Metaphilosophy* 36 (5): 634–651.

National Human Genome Research Institute: www.genome.gov.

Polanyi, Michael. 1951. *Logic of Liberty*. London: Routledge.

Resnik, David B. 2007. *The Price of Truth: How Money Affects the Norms of Science*. New York: Oxford University Press.

Chapter 5
Chemical Industry

Abstract Scientific research leading to the production of chemical agents and technologies enables malevolent agents to engage in harmful behaviour by way of a number of different pathways. For example, scientific research led to knowledge-how to aerolize chemicals for crop dusting (benefit); yet this discovery also made possible the aerolizing of chemicals for use in weaponry (harm). For nation-states (especially) can and do directly establish chemical weapons research programs. However, according to the definition in this book weapons research programs, including chemical weapons research programs, are not dual use because weapons are designed in the first instance to cause harm, and this is the case even if the weapons in question are developed for defensive rather than offensive use. Naturally, at least in principle, chemical weapons research might be conducted not with the intention of making and potentially using chemical weapons, but rather with the intention merely of understanding the functioning of such weapons so as to enable (say) the design and production of protective clothing in case of a chemical weapons attack by one's enemies. Such weapons research might be dual use in our favoured sense. It is also argued in this chapter that the management of dual use risks in chemical research should be seen as a collective moral and institutional responsibility of multiple actors that can only be fulfilled with a *web of prevention* (an integrated suite of regulatory measures).

5.1 Past, Present and Future Threats from Dual Use R&D in the Chemical Industry

Hitherto the literature on dual use problems has tended to focus on the biological sciences.[1] Dual use issues in the biological sciences are discussed in Chap. 8. In this chapter we consider an area of science and technology in which there are dual use

[1] See Note 1 Chap. 1.

This chapter was co-authored by Seumas Miller and Jonas Feltes.

© The Author(s) 2018   55
S. Miller, *Dual Use Science and Technology, Ethics and Weapons of Mass Destruction*, SpringerBriefs in Ethics, https://doi.org/10.1007/978-3-319-92606-3_5

problems that have thus far almost entirely escaped academic scrutiny, namely, the chemical industry. Two of the exceptions might be Tucker, and Becker and Trapp.[2] Furthermore, the collective moral (and associated institutional) responsibilities of the chemical industry that have arisen from dual use problems have not yet been subject to detailed academic investigation.

Scientific research leading to the production of chemical agents and technologies enables malevolent agents to engage in harmful behaviour by way of a number of different pathways. For example, scientific research led to knowledge-how to aerolize chemicals for crop dusting (benefit); yet this discovery also made possible the aerolizing of chemicals for use in weaponry (harm). However, nation-states (especially) also can directly establish chemical weapons research programs. As we saw in Chap. 2, the definition of dual use research and technology is contested and inevitably to some extent stipulative. However, according to the definition that we favour weapons research programs, including chemical weapons research programs, are not dual use because weapons are designed in the first instance to cause harm, and this is the case even if the weapons in question are developed for defensive use, i.e. even if the use of these weapons has as an ultimate purpose to achieve a benefit or good (in the sense of, for example, averting harm to members of one's own community).

Naturally, at least in principle, chemical weapons research might be conducted not with the intention of making and potentially using chemical weapons, but rather with the intention merely of understanding the functioning of such weapons so as to enable (say) the design and production of protective clothing in case of a chemical weapons attack by one's enemies. In some instances such weapons research might be dual use in our favoured sense. For on the one hand, the researchers' immediate and only end is to avert harm; there is no intention to harm others even as a means of self-defense. On the other hand, in some instances the research in question might potentially lead to knowledge of (say) more effective ways of weaponising chemicals, i.e. knowledge enabling others (secondary users) to weaponise chemicals more effectively for harmful purposes. These and similar issues of dual use in chemical research and development (R&D) shall be the subject of the present chapter. With this focus we do not only aim to shift the focus of ethical considerations in research to the chemical sciences, but also argue that the management of dual use risks in chemical research should be seen as a collective moral and institutional responsibility of multiple actors that can only be fulfilled with a multi-facetted *web of prevention*.

By contrast with weapons research, research on toxins might be undertaken in order to directly benefit rather than destroy or otherwise harm humankind, e.g. research into highly toxic pesticides. However, malevolent agents can steal chemical agents, specifically toxins, weaponise these toxins and, ultimately, perpetrate chemical attacks. Such attacks can be perpetrated using a multiplicity of weapons discharging chemical agents, including bombs, aerial bombs, rockets, artillery, tanks, landmines and spray mechanisms.

---

[2]Tucker (2012). Bakker and Trapp (2005, 13–17).

For our purposes here it is important to make a threefold distinction between lethal chemicals (e.g. sarin), incapacitating chemicals[3] (e.g. anaesthetic agents) and harassing chemical agents (e.g. CS gas used in riot control).[4] In this chapter lethal chemicals are of particular interest: The use of toxic chemicals as weapons occurred on a large scale in World War 1—notably mustard gas. According to Hay, the use of chemicals as weapons resulted in over million casualties in WW1 of whom 90,000 are estimated to have died.[5] For the most part chemical weapons were not used on a large scale for military purposes during World War 2. An exception was the Japanese use of chemical weapons in China.

However, although not frequently present in the battlefield, large quantities of chemical agents such as Zyklon B were used by Nazi Germany during WW2 to systematically murder millions of people in concentration camps. Originally developed as pesticides in the research laboratory of the chemist and Nobel prize winner Fitz Haber in the 1920's, the Zyklon agents and especially the later developed Zyklon B might be the most horrific examples of dual use in the history of chemistry.[6]

R&D into, and stockpiling of, chemical weapons went on during WW2 and in the second half of the twentieth century. Post WW2 R&D and stockpiling of chemical weapons reached a peak in the Cold War, during which both sides, the USA and the USSR conducted extensive and targeted military research to develop more and more deadly chemicals—a chemical arms race that led to the notorious V agents and (on the Soviet side) to the Novichok nerve agent that was designed to kill without detection.[7]

In recent times there have been various instances of chemical attacks perpetrated by nation-states. Saddam Hussein's regime in Iraq used mustard gas in the Iran/Iraq war (1980–88) and mustard gas against the Kurds (1988).[8] The Assad regime in Syria has been repeatedly using nerve agents such as sarin as well as chlorine gas and sulfur mustard against rebel groups as a recently published report to the UN shows in detail.[9] Furthermore, between January and September 2016, the Sudanese government allegedly used chemical agents in the remote area of Jebel Mara (Darfur)

---

[3]Note that in military terms sulfur mustard is not considered a lethal, but incapacitating agent since it is not lethal per se, but a blister agent. See Organisation for the Prohibition of Chemical Weapons (1992) art.II.9c.

[4]See World Health Organisation (2004).

[5]Hay (2005).

[6]Contrary to most journalistic and other popular accounts, the work of Haber himself cannot be linked directly to the development of the Zyklon agents. However, the work of his employees in the laboratory and especially the research and development of the Degesch company that was founded as a successor institution to Haber's laboratory was responsible for the development of Zyklon B. See for discussion Dunikowska and Turko (2011, 10050–10062).

[7]Although never used on a large scale, the Novichok agent has been used for targeted assassinations. At the time of writing the most recent example of such a case is the attempted assassination of Sergei and Yulia Skripal in Salisbury (UK) in March 2018. See Faulconbridge and Holden (2018). For general discussions concerning the Novichok agent see Sidell et al. (1997, 75).

[8]Szinicz (2005, 167–181 [172]); Bajgar et al. (2009, 17–24 [21–22]).

[9]OPCW (2016).

to fight the rebel group Sudan Liberation Army/Abdul Wahid as a report of the humanitarian organisation Amnesty international states. Although the nature of the agents remain unidentified, the symptoms of the (mostly civilian) victims suggest that the chemicals could include sulfur mustard.[10]

The use of chemical weapons has not been restricted to nation-states. Malevolent actors using chemical weapons, at least potentially, include 'lone wolf' individuals, nihilistic and apocalyptic ('end of the world') groups, and terrorists. They also include criminals who potentially, at least, could use the threat of a chemical attack for purposes of blackmail. The risks posed by such malevolent actors dramatically increase in unregulated spheres, such as in failed states and war zones and, of course, failed states parts of which are war zones, e.g. Syria at the time of writing. A relatively recent instance of a chemical attack by a nihilistic group was the sarin gas attack in a subway in Tokyo in 1995 perpetrated by Aum Shinrikyo. In this attack 12 persons were killed and five thousand others sought medical attention.[11] Terrorist groups, such as Al Qaeda and the self-proclaimed Islamic State of Iraq and Syria (ISIS) have also displayed an interest in the getting their hands on chemical weapons or, at least, the unmixed precursor chemical agents that would enable them to build chemical weapons; and there is every reason to believe that they would use chemical weapons if they possessed them. At the time of writing, the above mentioned report to the United Nations not only confirmed the use of chemical warfare against Syrian rebel groups by the Assad regime, but also pointed out that ISIS apparently used sulfur mustard in at least one occasion during the Syrian civil war.[12]

However, the weaponisation of toxic chemicals by malevolent agents might not be the only example for dual use issues in the chemical industry. Arguably, some industrial accidents might involve dual use issues, albeit this is not necessarily the case. In 1984 in Bhopal India at least 15,000 died (conservative estimate) and more than 200,000 were injured when toxic gas escaped from Union Carbide's insecticide plant. Those who suffered most were slum-dwellers living in the vicinity of the plant.[13] Bhopal was the world's worst industrial accident, worse than Chernobyl in terms of death and injuries. Responsibility for the disaster was collective in the sense that it was negligence on the part of Indian managers and employees at Bhopal—safety procedures specified in the handbook were routinely ignored, albeit by undertrained staff—and Bhopal's management in the US had not put in place an adequate accountability scheme. After the disaster Union Carbide tried strenuously to pay out as little as possible. It claimed employee sabotage—a somewhat farfetched claim—and wanted the matter to be tried in India where leniency and lesser payouts would be forthcoming. In the end a US judge determined that the matter be tried in India but under US legal principles. Union Carbide paid out $470 million and the Indian government agreed to drop criminal charges against Union Carbide.

---

[10] Amnesty International (2016).

[11] Danzig et al. (2011).

[12] Organisation for the Prohibition of Chemical Weapons (2016).

[13] Varma and Varma(2005, 37–45 [37–38]) and Cassels(1991, 1–50).

The Bhopal disaster illustrated the large-scale harm that can result when large quantities of chemical toxins are released into the atmosphere, deliberately or accidentally. However, those killed or injured were living in the vicinity of the plant and the safety procedures were woefully inadequate. The question that needs to be asked is whether the research (or R&D) that enabled the production of the toxic gas was dual use. While the toxic gas needed to be produced in large quantities if it was to result in large-scale harm, the potential for large scale harm, either by way of weaponisation or culpable negligence, clearly existed. Accordingly, it might reasonably be claimed that the research in this specific case was dual use. This is, of course, not to say that the research was not morally justified. After all, the researchers might have reasonably believed that the benefits outweighed the risks, on the assumption that reasonable safety precautions were taken. Accordingly, the conclusion to be drawn is that it was morally justified dual use research and the moral responsibility for the Bhopal disaster relies squarely on the managers of the Bhopal plant (both in the US and in India) rather than on the chemists who researched and developed the insecticide.

5.2   Dual Use R&D and the Chemical Industry

As we saw in Chap. 2, in the chemical industry the dual use problem arises, for example, in relation to research into pesticides. On the one hand, R& D into highly toxic pesticides, such as sarin-based pesticides, enables the eradication of pests which destroy crops. On the other hand, such R&D enables the production of the nerve agent, sarin, which can be used by the likes of Bashar al-Assad in the civil war in Syria to kill innocent civilians.

In relation to waging war by unacceptable means, we note that sarin produces uncontrollable nerve cell excitation and muscle contraction leading to death by suffocation. Hence it and other highly toxic chemical agents are outlawed under the Chemical Weapons Convention (CWC). The enforcement of this convention lies in the hand of the UN affiliated Organisation for the Prohibition of Chemical Weapons (OPCW) in The Hague.

However, the initial R&D into sarin and other pesticides was undertaken with good ends such as improving effectiveness in agriculture and, thereby, to counter undersupply, famines and hunger in societies around the world. Only secondary users, such as the rogue states or terrorist groups mentioned above, transformed the beneficiary innovation of sarin into a deadly weapon of mass destruction. Hence, the researcher conducting R&D into sarin and other nerve agents for good ends faces an ethical dilemma resulting from the secondary, potential harmful use of these innovations.

Understandably, governments have caused R&D to be undertaken on these toxins in order to protect members of their armed forces, and their citizenry, from chemical weapons. However, in some cases this has led to the development of even more toxic chemical agents. Such R&D is, therefore, dual use in character. It has been undertaken in order to provide a protection (good end) but has also resulted in the development of new and potentially more harmful toxins.

Notwithstanding the potentially harmful outcome of such dual use R&D undertaken for protective purposes, we need to distinguish it from R&D undertaken simply for offensive purposes. For example, the R&D into Novichok nerve agents by the Soviet Union was undertaken so as to render these nerve agents undetectable by NATO forces and, thereby, defeat their protective gear. Accordingly, this was not dual use R&D.

In summation, R&D in the chemical industry is for the most part undertaken with the purpose of providing a benefit, e.g. R&D on pesticides. The benefits in question include that of protection from chemical attack. In such cases, the original researcher does not undertake the research with any intention to harm; quite the reverse—s/he intends to benefit humankind. Nevertheless, in some cases there is the potential for a secondary user (e.g. terrorist, rogue nation-state, 'lone wolf') to intentionally use this R&D to harm, and do so on a massive scale. Hence the R&D in question is dual use R&D.

Thus far in this section we have assumed that the actual or potential harm caused by dual use R&D is intended (and intended by the secondary user). However, as we saw above in relation to the Bhopal disaster, there are some cases of dual use where harm potentially caused was not intended but, nevertheless, the (secondary) user is morally, if not legally, culpable. Negligence with respect to safety procedures at a plant producing toxic chemicals is one kind of case, but there are others. Consider, for example, the dumping of obsolete chemical warfare materials in the sea. This is an instance of failing to take adequate safety measures in the face of foreseeable harms, albeit unintended harms. Offenders in this regard include the USSR (undocumented dumpsites) and the US (post WW2 US dumping of obsolete chemical warfare materials in sea).[14] The potential harms arising from dumping chemical weapons at sea result from dredging, laying cables, and fishing. The harms include injuries to, or the deaths of, fishermen, contamination of fish (a public health risk), and environmental degradation (e.g. from damage to marine life ecosystems).

Granted that those who dump obsolete chemical weapons are culpable, what of the original researcher? Is the original researcher culpable by virtue of the actions (or omissions) of the secondary user in these kinds of case? Note that the secondary user is not using the original R&D as a weapon. Accordingly, these kinds of case are not instances of dual use R&D in this obvious sense. Nevertheless, just as the R&D of large quantities of chemical toxins is inherently dangerous because it may result in large-scale harm—as the Bhopal disaster graphically illustrated—so are chemical weapons capable of producing large-scale harm, even when not used as weapons but merely dumped at sea. On the other hand, R&D into chemical weapons is not dual use since the original researchers have as their immediate purpose the production of a weapon for harmful use rather than some peaceful purpose. (As mentioned above, this is the case even if their ultimate purpose is a peaceful one, e.g. to avert harm to themselves as in wars of self-defense.) Therefore, the harm caused by chemical

---

[14]See for discussion (ed) Kaffka (1996), and Newman and Verdugo (2010, 45–54).

weapons, whether by means of their use as weapons or by virtue of their being dumped at sea (or the like), is not dual use harm.

Some chemicals used for medicinal purposes can also be readily transformed and deliberately used for harmful purposes other than chemical warfare. Thus pseudoephedrine is used in decongestants. However, it is the precursor drug in methamphetamine or "crystal meth". Crystal meth is addictive and may lead to violent action, strokes and even death, not to mention the criminal activity undertaken to support a drug habit. Nevertheless, we suggest that such medicinal chemicals readily transformable into harmful drugs are not dual use chemicals. For the harm that might result is self-inflicted. It is not inflicted by others and it is not the result of culpable negligence by others with respect to an inherently dangerous chemical. We suggest that to widen the definition of dual use science and technology so that it included R&D implicated in such cases of self-inflicted harm would not be helpful. While scientists must take some responsibility for some of the potentially harmful uses of their research, they are not (so to speak) their 'brothers' keeper'.

As mentioned in Chap. 2, we further need to distinguish dual use R&D from R&D into harmful chemicals undertaken simply for offensive purposes. For example, the R&D into Novichok nerve agents by the Soviet Union was undertaken so as to render these nerve agents undetectable by NATO forces and, thereby, defeat their protective gear. Accordingly, this was not dual use R&D, since there was only one use intended in undertaking this R&D, namely to harm or kill persons. Secondly, as we saw above, chemical weapons developed only for defensive use (i.e. to be used offensively but only if attacked) are not instances of dual use R&D either. This is the case even if the R&D in question is, as some might argue, morally justified. Thirdly, R&D in which the actual or potential secondary user does not intend to use the R&D in order to conduct chemical attacks is not necessarily dual use R&D, although in some cases it might be. In cases in which R&D conducted for peaceful purposes, nevertheless, results in the manufacture of large quantities of an extremely harmful toxin, then this R&D may well be dual use as the example of the accident in Bhopal shows. On the other hand, since R&D conducted in order to produce chemical weapons is not per se dual use, then harms caused by this R&D are not dual use harms. For, as we have seen, harms caused by chemical weapons by using them as weapons are not dual use harms. Nor, as we have also seen, are harms caused by chemical weapons that are *not* being used as weapons dual use harms. Hence harms caused by dumping chemical weapons at sea are not dual use harms.

Of course, in the case of much actual chemical research (especially chemical research undertaken in the early twentieth century), the border between dual use research and chemical weapons research was very unclear; sometimes this unclarity was evident within one research institution: Although Degesch, the company that evolved out of the research laboratory of Fritz Haber, was conducting dual use research by investigating early versions of the Zyklon pesticides, it also conducted state funded research with the explicit aim to develop chemical weapons in the

1920s.[15] By contrast, as we have seen above, R&D in relation to weapons which are not to be used either for offence or defence, but rather to enable protection against attack using these weapons by others may well be dual use R&D. An example from the chemical industry is R&D into protective equipment against nerve agent attack which involves the R&D into the nerve agents in question. However, the problem here is the difficulty of determining that the R&D in question is, in fact, being undertaken for protective purposes, as opposed to for the purpose of constructing a weapon to be used for offence or defence. Thus it may be difficult in practice to determine whether an R&D program is engaged in developing a highly toxic agent, such as V agents or weaponry that uses V agents, merely in order to test its effects and enable detection and protection rather than in order to possess a weapon to be used in offence or defence. This thin line between research on protection and weapon focused research becomes especially visible by investigating the chemical arms race between the USA and the USSR during the Cold War.

Thus far we have talked in general terms of the harm implicated in dual use R&D. We now need to explicitly specify this harm somewhat utilising the definition of dual use harm outlined in Chap. 2. First, the harm in question is not dependent upon repetitive use. Thus an artefact such as, for example, a knife is not dual use technology merely because it could potentially over a long period of time be used to kill hundreds, even thousands, of people. Rather a one-off use of dual use technology can do great harm. Second, the kind of harm done by the use of dual use technology is in and of itself very serious harm, e.g. causing death, and the mode of harming is novel, e.g. by means of nerve agents. Thirdly, the harm done is on a very large scale, e.g. dual use R&D generates, actually or potentially, weapons of mass destruction (WMDs). All these conditions of harmfulness apply to weaponised highly toxic chemical agents. However, they evidently also apply to some man-made highly toxic chemical agents which are not weaponised, but whose dispersion into a human population (as a consequence, for example, of culpable negligence with respect to safety procedures) might cause large-scale harm.

As mentioned in the characterization of chemical agents above, some chemicals are harmful in the sense that they are incapacitating agents. In some cases these chemicals are also lethal. For example, Russian security forces used an incapacitating agent that was pumped into a theatre in Moscow in 2002 in order to free 800 hostages held by Chechen separatists. The Chechens were incapacitated and the hostages were released. However, 120 of the hostages died as a consequent of inhaling the chemical agent.[16] Indeed, according to the British Medical Association, current incapacitating chemical agents have a risk of death.[17] However, let us assume that non-lethal incapacitating chemical agents exist or are developed. An argument might be made that these are a useful tool in the hands of, for example, law enforcement agencies, since they incapacitate offenders without killing or seriously harming them. If so,

---

[15]Dunikowska and Turko (2011, 10050–10062); Manchester (2002, 64–69); Jansen (2000, 28–33 [31]).

[16]Crowley (2013).

[17]ibid.

then these chemicals do not give rise to dual use problems; for the secondary users are not engaged in deliberately (or otherwise culpably) causing harm on a massive scale, albeit the cumulative effect of each harming him/herself might be massive as seen in Moscow.

5.3   Individual and Collective Moral and Institutional Responsibility

In instances of dual use science and technology there is potentially a need for researchers to choose between the options of undertaking or not undertaking the research since they knowingly provide the means for the harmful actions of others. The researchers who develop V agents, for example, cannot simply absolve themselves of any responsibility for the deliberate harm done by others who weaponise these highly toxic chemical agents and engage in chemical attacks.

Naturally, these researchers are not directly responsible for such harm; they did not do the harming and they did not intend that others do the harming. Nevertheless, as we saw in Chap. 2, there is a relevant related principle, namely, do not provide others with the means to do large scale, serious harm, if one can avoid it: the 'No Means to Harm' principle or NMH. According to NMH, one ought not foreseeably (intentionally or unintentionally) provide others with the means to do serious harm on a large scale. In the paradigm case it is assumed that the toxin in question constitutes a means to do harm and others will, or may well, deliberately or knowingly do harm on a large scale, given the chance (or, at least, may act with culpable negligence).

It may also be the case that original researchers have a degree of culpability if; (i) they avoidably and foreseeably provide secondary users with the means to *accidentally* (and therefore, let us assume, non-culpably) do great harm, and; (ii) they know that it is *highly likely* that the secondary users will do great harm, albeit accidentally. However, even if so, the research in question is not necessarily dual use research in our favored sense.

Accordingly, it is important to note that the formulation of NMH principle set forth in Chap. 2 includes the condition that the harm needs to be of a very great magnitude. However, even thus understood the principle is not necessarily an absolute principle. For example, arguably it may be overridden in the case of R&D that has the potential to lead to the construction of chemical WMDs to be used as a deterrence by a liberal democratic state confronting the threat of chemical WMDs at the hands of a rogue state.

NMH principle is related to, but not identical with, a weaker moral principle: Act to Prevent Harm. Doubtless, one ought to act to prevent serious widespread harm, if one can. However, one is not necessarily implicated in such harm, if one does not act to prevent it. By contrast, (other things being equal, e.g. it is not a case of self-inflicted harm) one is implicated in harm done by others if one has knowingly provided them with the means to do harm. This distinction is crucial for the dual use R&D debate, since it, for example, directly addresses the chemist performing research on insecticides by, for instance, implicating her in harmful misuses of her work.

There is considerable complexity in the application of the NMH principle and of related principles, such as the principles of necessity and proportionality, due to the uncertainty of the future harms that might result from R&D. Moreover, this uncertainty might be thought to trigger the application of additional principles in the decision-making process, such as the so-called precautionary principle.[18] Accordingly, decision-making in relation to dual use R&D often involves complex evidential considerations, as well as moral ones, and, as such, is a matter for empirically and morally informed *judgment*, as opposed to mechanical application of clear-cut rules.

Irrespective of the particular principles that might need to be invoked, there are a number of general assumptions involved in any decision-making process in relation to dual use R&D that should be stressed (again). These include: (i) The potential harms are serious and on a large scale; (ii) the researcher is able—acting jointly with others—to prevent or reduce the harm by, for example, not engaging in the R&D; (iii) The social, economic, and moral costs of preventative measures are not excessive.

Accordingly, there are various individual and collective moral responsibilities derivable at least in part from NMH principle. Some of these are obvious but not dual use responsibilities such as, for example, selling chemical weapons to rogue states (as Frans van Anraat did[19]) or participating as a chemist in a national chemical weapons R&D program (as in the case of the Iraqi program under Saddam Hussein). With respect to dual use based responsibilities, some are individual and some are collective; moreover some are institutional and moral, others only moral (albeit, they might be such that they ought also to be institutional).

In the chemical industry, as elsewhere, dual-use challenges are not simply a moral problem for individual researchers/chemists. They are a problem for many other individuals, groups and organisations. Accordingly, there are multiple actors with moral responsibilities in relation to dual use problems. These include the following ones:

Governments: Members of governments have an institutional and collective moral responsibility to protect their citizens from the massive harm that might result from dual use R&D, e.g. by implementing appropriate regulatory measures in relation to R&D of toxins, export of toxins, destruction of chemical weapons etc.

Global community of nation-states: Members of governments and international bodies, such as the United Nations, have a collective institutional and collective moral responsibility to design and implement dual use regulatory arrangements, and to ratify and comply with the CWC and otherwise support the OPCW in its endeavours to eradicate chemical weapons and to educate about the dangers of these weapons.

Citizenry: Citizens have a collective responsibility to assist in their own protection e.g. by making themselves aware of dual use issues.

---

[18]Clarke (2005, 121–126).

[19]Tabassi and Van der Borght (2006, 36–44 [8–13]).

Military forces: Members of military forces have a collective institutional and collective moral responsibility not to use chemical weapons and otherwise to comply with the CWC.

Chemical industry firms: Members of the chemical industry have a collective moral responsibility to implement industry-wide dual use processes.

Chemical industry associations: The members of chemical industry associations have a collective moral responsibility to design appropriate dual use processes for their members.

Research organisations: Members of universities, government research bodies and commercial firms have a collective institutional and collective moral responsibility to implement appropriate dual-use processes.

All of the above individuals and organisations are part of the overall solution to the harms consequent upon dual use R&D in the chemical sciences. Moreover, the harms consequent upon dual use R&D are but component of the overall actual and potential harm resulting from R&D in the chemical sciences. Accordingly, the problem of dual use harm prevention needs to be framed as a fragment of the overall *collective* moral responsibility to prevent harm resulting from R&D in the chemical sciences. The collective moral responsibility in question is to design and implement an institutionally-based *web of prevention*.[20]

5.4 Collective Responsibility and the Web of Prevention

According to our favoured account of dual use R&D, dual use R&D in the chemical sciences has the potential for secondary users to deliberately cause large scale harm by deploying weaponised chemical agents. Therefore, prevention of such harm relies in part on the legal prohibition of the chemical WMDs themselves, as well as by introducing dual use regulation. As we have seen, the principal legal instrument for prohibiting chemical WMDs is the Chemical Weapons Convention. This treaty came into existence in 1992 and was subsequently enforced by almost all nation states via signing and ratifying the treaty or via ratification through accession (Israel is one exception since it signed, but never ratified the CWC[21]). The treaty prohibits the development, production, transfer and stockpiling of chemical weapons and requires signatories to destroy their existing stockpiles of chemical agents used in chemical weapons (above certain amounts), as well as the weapons themselves. The CWC provides lists of chemicals used in chemical weapons, or that could be converted for such use. However, the CWC does not prohibit the use of these chemicals; most are needed for peaceful purposes, e.g. in mining, as dyes, for medicinal purposes. Article II.9 of the CWC states under purposes not prohibited: "(c) Military purposes not connected with the use of chemical weapons and not dependent on the use of the toxic properties of chemicals as a method of warfare" (Organisation for the Prohibition

---

[20] See, for instance, Rappert and McLeish (2014).

[21] See for discussion Cohen (2001, 27–53).

of Chemical Weapons 1992 art.II.9c). The CWC provides for the monitoring of the production, transfer, stockpiling and use of its 'scheduled' chemical agents, and carries out inspections to ensure that the quantities of these chemicals do not exceed certain amounts. Importantly, the CWC has stringent verification procedures. In this respect it is unlike the Biological Weapons Convention (BWC).[22]

The Organisation for the Prohibition of Chemical Weapons (OPCW) is an independent body set up in order to give effect to the CWC. The OPCW monitors the production and stockpiling of chemical weapons (and the chemicals used in chemical weapons), carries out inspections, oversees destruction of stockpiles and publically communicates its findings. The Australia Group complements the OPCW. It comprises 39 nations and the European Commission and has similar objectives to the CWC. Its focus is on export licensing arrangements and it seeks to prevent the distribution of materials and equipment used in chemical weapons programs.

The CWC is a global institutional embodiment of the collective moral responsibility of nation-states to prevent chemical warfare. Each individual nation-state has a moral and institutional responsibility to comply with the provisions of the CWC interdependently with the discharging by the other signatory nation-states with their own like individual responsibilities.

As is the case with other global problems, eliminating chemical weapons confronts a collective action problem (see Chap. 4, Sect. 4.4). Each nation-state may well be prepared to forego chemical weapons if other nation-states do so, but what if the others do not? Here there is a collective good, namely, the elimination of chemical weapons by all, yet each only wants to eliminate its chemical weapons if the others do, and some might seek to continue to possess chemical weapons even if all others have eliminated them. So the problem is to get the nation-states to pursue this collective good; that is, to get a collective good to also be a *collective end*. An example for this collective action problem could be Israel's decision not to ratify the convention that appears to be, amongst others, grounded in the concern that its neighbor states either do not comply with the convention at all or would not be sanctioned if breaching it (the case of the Assad regime in Syria might be a recent example here).[23] The solution appears to lie in a strategy of incremental progress combined with stringent verification measures at each incremental step of the way. In the first stage, each nation-state agrees not to be first user of its chemical weapons, but each is allowed to continue to develop, produce and stockpile weapons. At the second stage, each agrees not to develop and produce new chemical weapons, and to reduce its existing stockpile. This step also brings with it an additional agreement to have one's claimed compliance with agreed reductions in chemical weapons subjected to scrutiny by the OPCW. Thus total prohibition becomes possible in the context of incremental progress. Indeed, this strategy of incremental progress is the one that has been implemented by the international community of nation-states and thus far it has been very successful,

---

[22]The difficulties and uncertainties surrounding verification were highlighted in relation to Saddam Hussein's biological and chemical warfare programs which existed but were then dismantled under United Nations supervision only to be said not to have been.

[23]See for general discussion Cohen (2001); Eitan (2010, 57–62).

even if not entirely successful (as the use by the Assad regime in Syria at the time of writing demonstrates).

That said, there remain some problems. For instance, and notwithstanding the progress made, there is a residual individual national interest that is apparently at odds with humanity's collective interest. An individual nation-state might reasonably believe that it ought to retain a small national chemical weapons program in order to understand chemical weapons so as to protect itself against them in the event that the CWC is imperfectly realized, e.g. a rogue state refuses to comply with the CWC or a nation-state is able to mask its chemical weapons program notwithstanding the application of verification procedures. Moreover, there is the very real possibility of non-state actors, such as terrorist groups, developing and producing chemical weapons or, at least, stealing them. Notice that a nation-state arguing in this manner for retention of a small chemical weapons program for protective purpose is hostage to the dual use dilemma. For its R&D undertaken for protective purpose may lead, for example, to the discovery of more toxic chemicals. At any rate, if this argument from residual national interest is correct, then although the strategy of incremental progress will go a long way to eliminating chemical weapons, it may well stumble at the final hurdle; and in doing so give rise to dual use problems.

More generally, solving the collective action problem of eliminating chemical weapons, supposing it occurs, might well turn out to be less important than might have initially been thought. For nation-states might well abandon chemical weapons, if they possess nuclear weapons. If so, the collective action problem of eliminating chemical weapons has in large part transmogrified into the collective action problem of eliminating nuclear weapons. This is, of course, not to suggest that eliminating chemical weapons is not a very good thing. But it is to downplay its importance in the overall context of eliminating WMDs and, more specifically, the resolution of dual use problems.

The signatories to the CWC are nation-states and the principal focus of the CWC is on the chemical weapons programs of nation-states and on what nation-states can otherwise do to eliminate chemical weapons. What of terrorist groups? Specifically, what of global jihadist terrorist groups, such as Al Qaeda and ISIS? These terrorist groups are unlike 'traditional' terrorist groups in a number of respects that are relevant to our current discussion. Firstly, their aims and reach are global; their terrorist activities are not confined to a limited geographical area or limited by a narrow political cause, such as was the case with groups such as the IRA or the Tamil Tigers. Secondly, they are far less discriminating in their choice of targets and far less restrained in their use of terrorist methods. Indeed, both Al Qaeda and ISIS have manifested a desire to acquire and use WMDs, something which most terrorist groups such as the IRA, the Red Brigades and the likes eschewed. This stems from the observation that especially members of ISIS present a worrying, apocalyptic ideology according to which the fight of the respective group is perceived as an end time battle against the infidels before the Day of Judgement.[24] Large scale attacks with biological and espe-

---

[24]See de Graaff (2016, 96–103).

cially chemical warfare would perfectly fit both this apocalyptic ideology and the respective descriptions of the end time in Islamic and Judeo-Christian eschatologies.

Thirdly, from a resource oriented point of view, developments in chemical R&D, including new manufacturing methods and associated micro-factories, have increased the potential for large-scale chemical attacks by terrorist groups and other malevolent actors. Furthermore, terrorist groups like ISIS are very likely to acquire already weaponised chemical agents in the environment of the Syrian civil war as shown in the above-mentioned UN report.

In addition to the availability of chemical agents, terrorists are also more likely to engage in chemical attacks than nuclear or biological attacks, since it is far more difficult for terrorists to acquire, let alone develop, nuclear weapons than it is for them to acquire or develop chemical weapons. And it is easier for attackers to direct and control chemical attacks than biological attacks; the latter are likely to lead to the infection of the terrorists themselves and of their supporters. In short, there is a far from remote possibility of a terrorist 'cottage industry' of weaponising chemical agents and subsequent use of these weapons. These developments in the potential technological wherewithal of terrorist groups are to some extent offset by developments in surveillance, detection and protection technologies. Moreover, the convergence of chemical and biological technologies has complicated matters. Nevertheless, the threat of the use of chemical weapons by terrorist groups is evidently increasing, all things considered.[25]

We suggested above that the responsibility to engage in dual use harm prevention in respect of R&D in the chemical sciences is a collective moral responsibility and, in particular, a collective moral responsibility to design and implement an institutionally-based *web of prevention*. As we have argued, this web of prevention includes the CWC, albeit the CWC is not specifically focussed on dual use problems per se. Indeed, this web of prevention needs to be designed in a manner such that dual use based-harm is only one source of the overall harm to be prevented. What are some of the other main parts of this web—in so far as the web is to address dual use based-harm?

In addition to the OPCW and the Australia Group, there is a need for regulatory authorities at national, industry and organisational levels. At the national level, the provisions of the CWC need to be enforced against individual persons and non-state organisations within the jurisdiction of each nation-state. Moreover, cross-border breaches of the CWC and related agreements, such as export of chemical weapons, by individual persons and non-state organisations should be criminalised under international law. In short, there ought to be criminalisation of breaches of the CWC by individual persons and non-state organisations.

Amongst other things, international, national, industry, occupational and organisational authorities (or, perhaps in some cases at the organisational and/or occupational levels, advisory committees) should assist in the process of identifying dual use problems in R&D and make adjudications/provide advice, both with respect to actually

---

[25]Discussion concerning the CBRN capabilities of groups like ISIS can be found in Ackerman and Pereira (2014, 27–34); House (2016, 68–75) and Hummel (2016, 18–21).

undertaking the dual use R&D in question and with respect to the dissemination thereof, e.g. in the mass media.

The regulatory architecture ought to include restrictions on stockpiles and export of toxins, prescribing of the safety and security-based conditions under which R&D can be undertaken and by whom, e.g. background checks and security clearance for research personnel, training programs, licensing of organisations, and so on. More-over, there is a need for educational and training programs that include material on dual use concerns. Ethics codes and codes of conduct are an important element of such programs. These codes can operate at an organisational and industry-wide level, but also at an occupational level. Thus different occupational groups, e.g. chemical engineers, chemists, might have their own codes, both at a national and interna-tional level. Further, elements of the regulatory architecture include protections for professional reporting of misuse of toxins and the like.

More generally, there is a need to restrict or, at least, stem the flow of knowledge in respect of the manufacture of highly toxic chemicals. The knowledge in question is collective expert knowledge (at least in the first instance) and it should, as far as possible, be restricted to relevant responsible expert sub-groups and certainly not allowed to become collective public knowledge. In short, there is a collective moral responsibility to maintain, where possible, collective public ignorance with respect to the manufacture of highly toxic chemicals and, specifically, curtail transfer of collective expert knowledge of these chemicals to malevolent actors. The justification for such knowledge restriction or censorship is well-known. Among other things, it is argued in effect that there is no general moral right to know how to produce WMDs or materials that might potentially cause serious, large-scale harm.

However, it is often responded to this kind of policy prescription that it is too late, that such knowledge is already 'on the internet'. In some cases this might be so, in other cases perhaps not. At any rate, the first general point to be made is that the purpose behind such restrictions is to reduce the risks, not eliminate the risks entirely. For the latter is not possible. The second point is that such restrictions need to work hand-in-glove with other measures and the security task might be less onerous in the context of the restrictions in question. The measures in question include not only regulatory ones, but also the acquisition of new knowledge that might assist by providing the means to protect against the harmful effects of the chemical toxins in question.

Specific issues that might need to be addressed include new developments in R&D, such as the convergence between chemical and biological technology, and the identification and resolution of collection action problems in otherwise well-regulated chemical industries. Thus firms in a given industry are in commercial competition. Firm X would strictly comply to the letter and spirit of the regulations concerning dual use R&D if X believed that all other firms did so. What if X does not believe this and that, therefore, X is at a commercial disadvantage? In such a situation in a highly competitive market, firm X (and, by analogy, firms Y, Z etc.) might make dual use cost/benefit judgments that favor commercial self-interest over strict compliance with, say, stringent safety and security regulations. Thus short term organisational commercial interests might override the long term public interest in ensuring dangerous toxins, for example, are not developed or, if developed, do not fall into the wrong hands.

5.5 Conclusion

In this chapter we have discussed the dual use problem as it manifests itself in the chemical industry. Chemicals are a necessary and ubiquitous features of modern society. However, there is a history of the same or related chemical agents being used in chemical weapons, notably in WW1. Moreover, chemical weapons are the most likely WMDs to be sought after by terrorist groups, such as ISIS—especially in light of recent developments in the Syrian civil war. A dual use concern of particular salience in the chemical industry is the development of protective equipment against chemical attacks using highly toxic agents as the case of V agents has shown.

We have argued that the responsibility to engage in dual use harm prevention in respect of R&D in the chemical sciences is a collective moral responsibility; specifically, a collective moral responsibility to design and implement an institutionally-based *web of prevention*. This web of prevention consists of the CWC and a range of additional regulations, governance arrangements and the like. These include restrictions on export of toxins, and prescribing of the safety and security-based conditions under which R&D can be undertaken and by whom, e.g. background checks and security clearance for research personnel, training programs, licensing of organisations, and so on. They might also include, where possible, measures designed to restrict the transfer of the knowledge required to create or manufacture particular toxins. After all, there is no general moral right to know how to produce WMDs or materials that might potentially cause large-scale harm. The recent reports from Syria and Darfur underpin this claim.

References

Ackerman, G.A., and Ryan Pereira. 2014. Jihadists and WMD: A Re-evaluation of the Future Threat. *CBRNe World* 2014: 27–34.

Amnesty International. 2016. *Scorched Earth, Poisoned Air. Sudanese government forces ravage Jebe Marra, Darfur*. London: Amnesty International.

Bajgar, Jiri et al. 2009. Global Impact of Chemical Warfare Agents Used before and after 1945. In *Handbook of Toxicology of Chemical Warfare Agents*, ed. Ramesh C. Gupta et al., 17–24 (21–22). Amsterdam: Elsevier.

Bakker, Edwin and Ralph Trapp. 2005. Chemicals—Good and Bad. In *Multiple Uses of Chemicals and Chemical Weapons*, 13–17. The Hague: OPCW.

Barak, Eitan. 2010. Getting the Middle East Holdouts to Join the CWC. *Bulletin of the Atomic Scientists* 66 (1): 57–62.

Cassels, J. 1991. The Uncertain Promise of Law: Lessons from Bhopal. *Osgoode Hall Law Journal* 29 (1): 1–50.

Clarke, Steve. 2005. Future technologies, dystopic futures and the precautionary principle. *Ethics and Information Technology* 7 (3): 121–126.

Cohen, Avner. 2001. Israel and chemical/biological weapons: History, Deterrence, and Arms Control. *The Nonproliferation Review* 8 (3): 27–53.

Crowley, M. 2013. Exploring the Role of Life Scientists in Combating the Misuse of Incapacitating Chemical and Toxin Agents. In *On the Dual Uses of Science and Ethics: Principles, Practices and Prospects*, ed. Brian Rappert, and Michael Selgelid, 293–330. Canberra: ANU Press.

Danzig, R., M. Sageman, T. Leighton, L. Hough, H. Yuki, R. Kotani and Z. M. Hosford. 2011. *Aum Shinrikyo. Insights into How Terrorists Develop Biological and Chemical Weapons*. CNAS Archive. https://s3.amazonaws.com/files.cnas.org/documents/CNAS_AumShinrikyo_Danzig_1.pdf?mtime=20160906080509.

de Graaff, Bob. 2016. IS and Its Predecessors: Violent Extremism in Historical Perspective. *Perspectives on Terrorism* 10 (5): 96–103.

Dunikowska, Magda, and Ludwig Turko. 2011. Fritz Haber: The Damned Scientist. *Angewandte Chemie International Edition* 50 (43): 10050–10062.

Faulconbridge, Guy and Michael Holden. 2018. Explainer: The poisoning of former Russian double agent Sergei Skripal. Reuters.com, March 13.

Hay, A. 2005. Toxicology of Chemical Warfare Agents. In: *Multiple Uses of Chemicals and Chemical Weapons*. The Hague: OPCW.

House, Carole N. 2016. The Chemical, Biological, Radiological, and Nuclear Terrorism Threat from the Islamic State. *Military Review* 96 (5): 68–75.

Hummel, Stephen. 2016. The Islamic State and WMD: Assessing the Future Threat. *CTC Sentinel* 9 (13): 18–21.

Jansen, Sarah. 2000. Chemical-Warfare Techniques for Insect Control: Insect "pests" in Germany before and after World War I. *Endeavour*, 24(1): 28–33 (p. 31).

Kaffka, Alexander V. (ed.). 1996. *Sea-Dumped Chemical Weapons: Aspects, Problems and Solutions*. Dordrecht: Springer Science & Business Media.

Manchester, Keith L. 2002. Man of Destiny: The Life and Work of Fritz Haber. *Endeavour* 26 (2): 64–69.

Newman, Joshua and Dawn Verdugo. 2010. Building Awareness of Sea-Dumped Chemical Weapons'. *Disarmament Forum Maritime Security*, 45–54.

Organisation for the Prohibition of Chemical Weapons. 1992. *Convention on the Prohibition of the Development, Production, Stockpiling and Use of Chemical Weapons and on Their Destruction*, art.II.9c. The Hague: OPCW.

Organisation for the Prohibition of Chemical Weapons. 2016. *Third report of the Organisation for the Prohibition of Chemical Weapons—United Nations Joint Investigative Mechanism 3/98 16-14788*. The Hague: OPCW.

Rappert, Brian, and Caitriona McLeish. 2014. *A Web of Prevention: Biological Weapons, Life Sciences and the Governance of Research*. New York: Routledge.

Sidell, F.R., E. T. Takafuji and D. R. Franz. 1997. *Medical aspects of chemical and biological warfare*, 75. DTIC Document.

Szinicz, L. 2005. History of Chemical and Biological Warfare Agents. *Toxicology* 214(3): 167–181 (172).

Tabassi, L. and E. van der Borght. 2006. Chemical Warfare as Genocide and Crimes Against Humanity. *CBW Conventions Bulletin*, 36–44 (8–13).

Tucker, Jonathan. (ed.). 2012. *Innovation, Dual Use, and Security: Managing the Risks of Emerging Biological and Chemical Technologies*. Harvard, MA: MIT Press.

Varma, Roli and Daya R. Varma. 2005. The Bhopal Disaster of 1984. *Bulletin of Science, Technology & Society* 25(1): 37–45 (37–38).

World Health Organisation. 2004. *Public Health Response to Biological and Chemical Weapons: WHO Guidance*. Geneva: WHO.

Chapter 6
Nuclear Industry

Abstract Dual use problem exists in an acute form in the nuclear sciences and technology. For scientific research, technology and materials in the nuclear sciences have enabled, on the one hand, unbounded nuclear energy for peaceful purposes, and yet on the other, massive arsenals of nuclear WMDs with the potential to destroy humankind. The nuclear industry has also facilitated the potential for malevolent actors to deploy 'dirty bombs' and created the conditions under which culpable negligence can result in nuclear disasters. Moreover, scientists and technologists have specific moral responsibilities, both individual and collective, given their central roles in nuclear power plants and government sponsored weapons programs, in particular. These include the collective responsibility to maintain collective public ignorance and not to enable collective expert knowledge in respect of certain expert groups. The generic solution to the kind of collective action problem to be found in the nuclear sciences, notably the nuclear weapons arms race, is at least in part an enforced cooperative scheme: enforced cooperative schemes are one important way to embed collective moral responsibility in institutional settings suffering from harm inducing collective action problems. This requires widening and strengthening existing institutional arrangements such as the NPT, but also creating additional ones, especially in the area of enforcement.

As noted in Chap. 5, in recent times the academic literature explicitly dealing with dual use problems has tended to focus on the biological sciences.[1] In Chap. 5 we focussed on the chemical industry. In this chapter we consider nuclear science and technology. In doing so we return to a debate that had currency circa mid-twentieth century in the wake of the Manhattan project and the development of the atomic bomb.[2]

---

[1] Miller and Selgelid (2007).
[2] Schweber (2000).

This chapter was co-authored by Seumas Miller and Behnam Taebi.

© The Author(s) 2018   73
S. Miller, *Dual Use Science and Technology, Ethics and Weapons of Mass Destruction*, SpringerBriefs in Ethics, https://doi.org/10.1007/978-3-319-92606-3_6

6.1 Dual Use Issues in Nuclear Science and Technology

World-wide the nuclear energy industry makes use of several hundred nuclear reactors and the numbers are increasing, notwithstanding the reversals for the nuclear industry in Germany and Japan following on the Fukushima-Daiichi nuclear reactor disaster in 2011.[3] Moreover, the US and Russian stockpiles of nuclear weapons comprises tens of thousands of warheads. Aside from any other moral consideration, stockpiles of this magnitude—and the inherent risks that they pose in terms of safety and security—are in clear violation of the principle of proportionality (as it might apply to arms build-ups, as opposed to actual use of arms). Clearly, nuclear technology poses very significant risks. Generally, we could distinguish between four types of nuclear or radiological risks,[4] namely: (i) detonation of an actual nuclear weapon; (ii) sabotage of a nuclear facility (such as a reactor); (iii) radiological dispersal devices or the so-called "dirty bombs"; and (iv) an accident with a nuclear facility (e.g. Chernobyl and Fukushima-Daiichi).

Naturally, some of these risks are not dual use risks per se, e.g. the risks posed by nuclear weapons programs or the unforeseeable accidents should not be counted as dual use. (See also the discussion in Chap. 2). Nevertheless, nuclear technology does pose a number of dual use risks. The nuclear energy use in Iran is a telling illustration of dual use risk, although it might be claimed that the original technologists and the (potential) malevolent secondary users will turn out to have the same political masters. While the nuclear energy facilities in Iran undoubtedly serve peaceful purposes, some technologies also have the potential to facilitate the production of nuclear weapons.

Dual use problems associated with nuclear technology are more familiar and more pronounced than in other fields. Perhaps this is in part because of nuclear disasters, such as Chernobyl in 1986 (involving the release of substantially more radioactive material than the atomic bomb dropped on Hiroshima) and Fukushima-Daiichi, and, indeed, in part because of the dropping of atomic bombs on Hiroshima and Nagasaki by the US in World War 2, and the subsequent threat of nuclear war—and, consequently, the potential for the obliteration of humankind—between the US and the Soviet Union during the post World War 2 Cold War period.[5]

At the heart of the dual use problem in the nuclear sciences and technology is nuclear fission. On the one hand, nuclear fission (the process of splitting of the nucleus of an atom) enables the generating of power and the creation of radioisotopes for use in, for example, the treatment of cancer. On the other hand, nuclear fission enabled the making of the atomic bombs that killed hundreds of thousands of Japanese. Moreover, the blasts from 'dirty bombs' embed radioactive particles in their human targets, and contaminate infrastructure with radioactive ash. While the production of nuclear weapons is likely to be beyond the capabilities of most, if not all, terrorist

---

[3] Taebi and Roeser (2015), 2.

[4] Bunn et al. (2016). Bunn et al. identified the first three categories of dual use risks.

[5] A notable episode in the decades long Cold War was the Cuban Missile Crisis of 1962. See, for example, Scott and Hughes (2015).

groups, the theft of radioisotopes and manufacture of 'dirty bombs' may well not be.[6] We note that a number of states with nuclear energy industries and, in some cases, nuclear weapons suffer ongoing and widespread terrorist attacks. The latter include India and Pakistan.

The processes that enable fuel to be made for nuclear reactors generating power for peaceful purposes can also enable the production of explosive material for nuclear bombs. Here there are two pathways. The uranium pathway involves the enrichment of natural uranium to facilitate effective fission. Low enriched uranium (around 3–5%) is sufficient for fuelling most existing power reactors generating electricity. Highly enriched uranium (up to 70–90%) is necessary for producing nuclear weapons; the Hiroshima bomb was based on highly enriched uranium. However, essentially the same technology as used for producing low enriched uranium (but in different scales and configurations) can also be used to make highly enriched uranium. This makes uranium enrichment technology a morally problematic technology.

The other pathway involves the plutonium that is produced in nuclear reactors when using uranium as fuel. Spent (or used) fuel coming out of the reactor can be considered waste and be allocated to final disposal places underground; this is called the once-through cycle and it is currently common in the US but also in several European countries like Sweden and Finland. Alternatively, spent fuel could be recycled or reprocessed. The unused uranium and the produced plutonium are then chemically separated from the irradiated fuel. In principle, when this is done with peaceful purposes in mind, both uranium and plutonium must be reinserted into the cycle (as the so-called Mixed Oxide Fuel); this nuclear power production method is called the closed fuel cycle and it is common in many European countries. Plutonium could, however, also be potentially used for manufacturing a nuclear explosive device. For instance, the Nagasaki bomb was a plutonium based bomb. It should, however, be noted that the plutonium that—under normal circumstances—would emerge from a conventional power reactor is not very suitable for weaponisation because of its low yield. This requires some explanation. When uranium oxide is irradiated, different materials including different *types* of plutonium (also called isotopes) would be produced. It is only one of those isotopes that would be most useable for weaponisation.[7] Nevertheless, it is important to include reprocessing as a dual use technology because, firstly, reactor-grade plutonium does possess some destructive power and could in principle be used for a (relatively low yield) device and, second, when a country possess such facilities it could in principle produce and extract suitable plutonium for weapon purposes (weapon-grade plutonium).[8] It is noteworthy that reprocessing plants were originally built for military purposes. It was only after Eisenhower's "Atoms for Peace" speech in the United Nations General Assembly in 1953 that

---

[6]Evans (2013), 262–3.

[7]For a detailed explanation of this phenomenon, see pp. 303–304 in Taebi (2012), 295–318.

[8]One way for manufacturing better useable plutonium would be using a rather short irradiation time of uranium fuel in a conventional reactor. Moreover, some reactors would be better equipped to produce the type of plutonium that is useable for weapon purposes; see Goldberg and Rosner (2011). Future reactors are being designed to also address (and reduce as much as possible) the risks associated with security and non-proliferation; see Taebi and Kloosterman (2015), 805–829.

reprocessing was considered a civilian technology too. The United States promoted reprocessing as one of the key technologies for efficient use of nuclear material for energy production for several decades afterwards.

Currently, only a handful of countries have reprocessing plants and the only non-nuclear weapons country that has a reprocessing plant is Japan. As Japan is a country with virtually no fossil fuel, it was planning—in the pre-Fukushima era—to use its imported nuclear fuel most efficiently; extensive and repeated reprocessing (in conjunction with nuclear breeder reactors) was part of the plan. In principle civilian reprocessing technology is proposed to, firstly, use the fuel more efficiently and, secondly, reduce the waste life-time. With dropping natural uranium prices very few countries insist on having access to reprocessing. While the waste life-time reduction is based on several (sometimes unsubstantiated) assumptions, many European countries have opted for the closed cycle and shipped their spent fuel to France and the UK to be reprocessed and shipped back.[9] The same goes for uranium enrichment that are only present in a handful of countries who—in principle—provide fuel for other nuclear energy producing countries.[10] Accordingly, in addition to uranium enrichment, plutonium reprocessing technologies are highly morally problematic too since they enable the production of nuclear weapons.

Some dual use problems arise without involving weaponisation on the part of secondary users; specifically, as we saw in Sect. 6.1, as a result of culpable negligence on the part of secondary users. Arguably, Chernobyl and Fukushima-Daiichi illustrate this point. These disasters, especially Chernobyl, illustrate the large-scale harm that can result when large quantities of radioactive material are released into the atmosphere, deliberately or accidentally. Evidently, the potential problem at Chernobyl, in particular, was reported prior to the disaster. The reports were ignored and even when the melt-down occurred there was an attempted cover-up and a delayed response. So the disaster was partly avoidable and the delay in response after it had happened exacerbated the situation. The question that needs to be asked is whether the research and/or technology that enabled the production of, or actually produced, the radioactive material was dual use. Presumably, the answer is in the affirmative, since the potential for large scale harm, either by way of weaponisation or culpable negligence with respect to man-made, extremely harmful, radioactive material, clearly existed. The fact that the research and/or technology were dual use is not, of course, to say that it was not morally justified, all things considered. After all, the scientists and technologists in question might have reasonably believed that the benefits outweighed the risks, and done so under the assumption, false as it turned out, that reasonable safety precautions would be taken. Perhaps the conclusion to be drawn is that it was a case of morally justified dual use research/technology and the moral responsibility for the Chernobyl disaster (and, for that matter, the Fukushima-Daiichi disaster) relies squarely on the relevant institutional actors who regulated

---

[9]For a detailed discussion of the mentioned assumptions upon which reprocessing is carried out, see Taebi (2013), 259–280.

[10]Also research reactors sometimes run on enriched uranium. So non-nuclear energy countries that have research reactors rely on such suppliers too.

and managed these plants (and their political masters, at least in the case of Chernobyl) rather than on the scientists and technologists whose R&D and subsequent scientific input enabled the existence of the nuclear plants and nuclear processes.[11] More generally, it might be argued that dual use research/technology in the nuclear sciences undertaken for peaceful purposes (notably, nuclear power plants to meet civilian energy needs) is morally justified, all things considered, given reasonable safety precautions are taken. Perhaps the most obvious problem with this line of thinking is the assumption of reasonable safety precautions, both at the level of an individual power plant and at the whole of industry level, in the historical context of the ongoing existence of a variety of malevolent and culpably negligent actors such as totalitarian states, failed states, terrorist organisations, and so on. Naturally, more detailed discussion is called for if this issue is to be resolved, including recourse to the specific risk assessment, decision-making procedures and accountability mechanisms in place. Moreover, we need to distinguish between the historically-focused dual use questions and the dual use questions that now confront us, given the reality of nuclear industries, nuclear weapons etc. It is the latter that are of primary importance, albeit the solutions depend in part on national and international regulatory and other institutional arrangements put in place over a long period of time.

Aside from dual use harms arising from nuclear R&D and nuclear accidents, there are those potentially arising from radiological dispersal devices made from radioisotopes (including the ones mentioned above), also known as 'dirty'. The recent terrorist attacks in Paris and Brussels serve to draw our attention to this kind of dual use problem. According to a report in the *International New York Times*,[12] "[t]he investigation into this week's [March 2016] deadly attacks in Brussels has prompted worries that the Islamic State is seeking to attack, infiltrate or sabotage nuclear installations or obtain nuclear or radioactive material. This is especially worrying in a country with a history of security lapses at its nuclear facilities, a weak intelligence apparatus and a deeply rooted terrorist network…. Asked on Thursday at a London think tank whether there was a danger of the Islamic State's obtaining a nuclear weapon, the British defence secretary, Michael Fallon, said that "was a new and emerging threat."

Historically, the prevention of the proliferation of nuclear technology has relied on the Non-Proliferation Treaty (NPT) and the Nuclear Suppliers Group (NSG). The NSG complements the NPT by limiting exports of nuclear materials and technologies, including dual use technologies. Under the NPT only the United States, Russia, China, the UK and France are recognised as nuclear weapon states because they produced and detonated nuclear weapons prior to 1967. Under article VI of the NPT, all states, including the nuclear weapons states, must make good faith efforts to achieve nuclear disarmament. Despite these powerful international agreements, there have been at least four new proliferators added to the list of nuclear weapons possessing countries since the signing of the NPT, namely, India, Pakistan, Israel

---

[11]It has been claimed that the regulators were not sufficiently independent in that they were not at a 'safe' distance from the nuclear industry.

[12]Rubin and Schreurer (2016).

and North-Korea. The first three have never signed the NPT, while Israel is believed to have had undeclared nuclear weapons before the NPT was even completed. When the NPT was signed and ratified it was projected that there would be a far larger number of nuclear weapon states by now; however, there has actually been no net increase in the number of nuclear weapon states for a quarter century (North Korea joined the group and South Africa left it).[13]

Moreover, a number of countries have openly or clandestinely pursued nuclear ambitions, either through a program dedicated to the development of nuclear weapons or through dual use nuclear technologies that are particularly troublesome. Under article IV of the NPT each nation-state has the right to peaceful nuclear technologies if but only if it has appropriate assurances of its peaceful nuclear program and does not manufacture nuclear weapons. However, the NPT does not explicitly exclude enrichment and processing technologies.[14] The controversies surrounding the Iranian nuclear programs vividly illustrate the complexities of this dual use of nuclear technology and the ambiguity of the NPT as an international treaty. While Iran keeps emphasizing its *inalienable right* to nuclear technology for civil purposes (Article 4, NPT), many countries dispute whether Iran should also develop uranium enrichment as a dual use nuclear technology. Under the Joint Comprehensive Plan of Action (JCPOA), also known as the Iran-deal, the country is going to keep its enrichment facility but under serious international monitoring for at least the next fifteen years; it should be mentioned that all declared enrichment facilities are in all times under the IAEA monitoring.

It is worth noting that the countries who have clearly contravened their treaty obligations were all non-democratic states at the time of the contravention. They include North Korea, Iraq and Syria.[15] Iraq's nuclear weapons program was terminated by US led coalition forces in 1991 shortly after Saddam Hussein's Iraqi forces invaded Kuwait.[16] India, Israel and Pakistan have nuclear weapons. However they are not signatories to the NPT. The theft in 1976 of plans for a centrifuge from Urenco in the Netherlands by Dr Abdul Qadeer Khan led to Pakistan's acquisition of nuclear weapons.[17] Nuclear trade restrictions on India under the NSG were lifted in 2008.[18]

The International Atomic Energy Agency (IAEA) provides essentially the safeguards system for the NPT. The IAEA conducts inspections of signatories to the NPT with the purpose of detecting and publically reporting any diversion of peaceful nuclear activities to military activity. The key elements of such diversion are the scientific knowledge and the fissile material required to build a nuclear weapon.

While the NPT and the IAEA perform important regulatory safety and safeguarding functions, however imperfectly, at the international level, what of the national level? While there is a wealth of international agreements and guidelines (mostly

---

[13]Taebi and Roeser (2015).

[14]Ferguson (2011).

[15]Miller and Sagan (2009), 11.

[16]Butler (2001), 83.

[17]Corera and Myers(2009).

[18]Meier and Hunger (2014), 17.

by the IAEA) in place,[19] hitherto managing nuclear safety and security within the nation-state has been in large part a matter of state sovereignty.[20] In the US, for example, nuclear R&D has from its inception been conducted under the doctrine of "born secret"—as expressed, for example, in the Atomic Energy Act of 1946—unlike, for example, R&D in the biological sciences. According to Howard Morland, before the Manhattan Project (to develop the US atomic bomb), US government secrets were temporary.[21] The Atomic Energy Act of 1946 had as one of its stated purposes to control all scientific and technical information concerning the manufacture of atomic weapons, the production of fissionable material for atomic weapons *and for the production of power*, unless it was declassified.[22] This highlights the importance of the issue of collective ignorance, especially collective public ignorance but also, in the case of some potential expert groups, collective expert ignorance (see Chap. 3, Sect. 3.2).[23] On the one hand, collective ignorance in these different senses assists in reducing the chances of nuclear proliferation. This is consistent with their being a degree of collective knowledge, collective public *propositional* (as opposed to practical) knowledge in particular, with respect to general features of nuclear processes, nuclear weaponry and the like. For the latter increases the chances of responsible decision-making by governments, and of input by citizens into decisions in respect of the nuclear industry. Accordingly, there is a need to determine what kind, level and mix of collective ignorance and collective knowledge is desirable among citizens, governments and experts.

6.2 Individual and Collective Moral Responsibility of Scientists

The nuclear industry is heavily regulated and closely controlled by governments; accordingly, nuclear weapons and nuclear power are in large part regulated and controlled by governments. That said, the scientific and technical knowledge that underpins nuclear weapons and nuclear power depends on scientists.

Wernher von Braun was a physicist and rocket scientist who developed the V2 rocket for Nazi Germany which was used against Britain in World War 2. After World War 2 he worked on rockets for the US. Evidently, von Braun had few moral scruples, notwithstanding his commitment to the scientific enterprise. Robert Oppenheimer worked to produce the atomic bomb that was used on Hiroshima and Nagasaki. Andrei Sakharov is known as the 'father' of the Soviet hydrogen bomb, albeit in his later life he worked to ban nuclear weapons. In neither case was this scientific

---

[19]Findlay (2011).

[20]While not fully successful, the IAEA does play a key role in in relation to verification of compliance with the NPT. See Findlay (2012).

[21]Morland (2005), 1401–8.

[22]Ibid. 1402.

[23]Miller (2017).

work dual use; rather it was *intended* to lead to serviceable weaponry and, in Oppenheimer's and Sakharov's cases, nuclear weaponry. Clearly, therefore, the scientist, Oppenheimer, bears some considerable moral responsibility (jointly with others), not only for the production of an atomic bomb, but for the tens of thousands killed when it was dropped on Hiroshima and Nagasaki. For his part, the scientist, Sakharov, clearly bears some considerable moral responsibility (jointly with others) not only for the production of a hydrogen bomb, but for the production of a hydrogen bomb for an authoritarian and expansionist state. I note that, arguably, authoritarian expansionist states are less likely to respect principles of necessity, proportionality etc. (i.e. principles of Just War Theory) in relation to waging war.

The cases of von Braun, Oppenheimer and Sakharov testify to the absurdity of the proposition that the scientific enterprise somehow stands outside economic, social and political institutions and purposes. These cases also testify to the willingness of scientists to subjugate their scientific activities to political and military purposes; science in the service of political, indeed military, power.

The larger point to be made here is that while science often provides the means it cannot determine the ultimate ends; specifically, the ultimate collective ends of human activity. This point holds, paradoxically, for science as an end-in-itself. For to choose science or, more broadly, understanding as an end-in-itself is not a *scientific* decision per se. This claim needs to be distinguished from the related one that truth is internal to science, as truth is to knowledge-seeking, more generally. After all, in making judgments one cannot aim at falsity, albeit one can assert, or otherwise communicate to others, what one knows to be false. However, to aim at the truth is not necessarily to aim at the truth for its own sake; to have something as an unavoidable end is not necessarily to have that thing as an end-in-itself. Moreover, many of these activities are not scientific (truth-seeking) discoveries per se; they are merely focussed on technological development that will facilitate the production of weaponry.

Further, as argued throughout this work, scientists are not exempt from the general moral obligation not to cause serious harm to others and from various derived obligations (and associated moral principles, e.g. those of necessity and proportionality). Specifically, as argued in Chap. 2, scientists have a derived moral obligation not to provide others with the means to do large scale, serious harm, if they can avoid it at relatively little cost to themselves; this (in simple terms) is the 'No Means to Harm' principle or NMH introduced in Chap. 2, Sect. 2.3. To recap: according to NMH, one ought not avoidably and foreseeably (intentionally or unintentionally) provide others with means to do serious harm on a large scale.[24] (As also mentioned, the application of NMH involves, in particular, consideration of the principles of necessity and proportionality.)

It is important to note here, as elsewhere, that the formulation of NMH is not necessarily an absolute principle. For example, arguably it may be overridden in the case of R&D that has the potential to lead to the construction of nuclear WMDs to be used merely as a deterrence by a non-aggressive, non-expansionist, liberal

---

[24]Miller (2013).

democratic state confronting the threat of nuclear WMDs in the hands of an aggressive authoritarian state with expansionist ambitions.

Given the embeddedness of R&D in the nuclear sciences in social, economic and political institutions, including private sector nuclear power plants, universities, and government sponsored weapons programs, and given also the NMH, nuclear scientists and technologists have individual and collective moral responsibilities with respect to dual use R&D in the nuclear sciences. These moral responsibilities pertain both to undertaking R&D and publishing the results of this R&D. Naturally, here as elsewhere, there are important differences in the nature and degree of the responsibilities in question. Scientists who improve the efficiency of the fission reaction, in principle to be used in the peaceful application but also applicable to the military use, may well not bear the same degree of responsibility for harm consequent upon future military use of weapons that relied on their work as ones who are designing a nuclear carrier device that resulted in the same extent of harm in a future military conflict.

Notwithstanding the *legal* obligation under the US Atomic Energy Act, scientists had a *moral* obligation to disclose to the public general information about the development of the hydrogen bomb, even if not detailed information that might have enabled malevolent members of the public or foreign powers to develop such a bomb. In the first place, the US is a democracy and its citizens, therefore, have a collective right (in the sense of a joint right) to collective knowledge (in some appropriately qualified form, such as public propositional knowledge (see Chap. 3, Sect. 3.1), which would exclude knowledge sufficient to enable the construction of nuclear weapons, for instance) about important programs and policies of its government, including for the purpose of voting for, or against, them. In the second place, the global community has a collective (i.e. joint) right to collective knowledge (again, in some appropriately qualified form) about nuclear weapons programs, in particular, given the potentially devastating effect they might have on the human race as a whole or, at least, on very large populations of civilians.

Moreover, these *moral* responsibilities of scientists and technologists may be inconsistent with their individual and collective *institutional* responsibilities. After all, the role occupants in nuclear power plants and nuclear weapons programs have *institutional* responsibilities. Sakharov, for example, had an institutional responsibility qua scientist in the Soviet Union's nuclear weapons program to assist in the development of the hydrogen bomb. But presumably he had no such *moral* responsibility; and certainly von Braun did not have a moral responsibility to assist the Nazi regime to develop the V2 rocket to be used against the civilian population of London. We note that such stringent moral responsibilities involving the potential for large-scale harm typically trump narrow institutional responsibilities when the two come into conflict.

It might be argued that the work of scientists and technologists in nuclear weapons programs in liberal democracies is morally defensible, notwithstanding that this is not so for authoritarian or totalitarian states, such as Nazi Germany or the Soviet Union under Stalin. The argument here is presumably based on the right to self-defence and, in particular, the right to deter a would-be attacker. Perhaps so. However, the right to self-defence, based on possession and use of offensive weapons, is not to be equated to the right to protect one's-self against an attack since the latter does not necessarily involve the use of offensive weapons. Moreover, only the latter potentially involves dual use R&D; R&D conducted for the purpose of making weapons for actual or potential use as weapons is not dual use R&D.

Further R&D in the nuclear sciences engaged in for purely protective purposes, such as the US' so-called Star Wars program initiated by President Reagan, is morally problematic and by virtue of its dual use character. For, as with dual use R&D in general, it may facilitate the production of ever more lethal nuclear weapons as a response to the new and emerging protective technology.

It might be thought that whereas participation by scientists in nuclear weapons programs is morally problematic—whether it be for protective purposes only or not—nevertheless, participation by scientists in the nuclear energy industry under-taken for peaceful purposes is morally *unproblematic*. However, as we have seen, quite central R&D in the nuclear sciences is dual use R&D; specifically, processes involving fission and the production of fissionable materials. We say this notwith-standing the fact that these processes are now well-known; so the cat is well and truly out of the bag (so to speak) as far as collective expert knowledge of these processes is concerned. Accordingly, the regulatory emphasis is rightly in large part focused on the presence of well-understood enrichment and reprocessing processes rather than on original scientific research per se. And, as we have seen, dual use science, technology and materials are inherently problematic. Indeed, dual use science, tech-nology and materials in the nuclear sciences even more so than in other sciences, given the potential of nuclear war and the immensely catastrophic consequences for both humankind and the environment.

6.3 Collective Action Problems in the Nuclear Industry

The threat posed by nuclear weapons is so great that its avoidance is, or ought to be, a collective end that overrides almost any other consideration. Accordingly, reasoning conducted by rational and moral agents would have the avoidance of nuclear war as a shared end. This is not to say that the elimination of nuclear weapons does not constitute a collective action problem in need of a non-obvious solution (see Chap. 4, Sect. 4.4). Moreover, there are other collective action problems that beset the nuclear industry, notably, given our concerns in this chapter, dual use problems, but also safety issues that might not be dual use problems. Collective action problems in the nuclear industry can be conveniently thought of as existing on a spectrum at one end of which is the problem of eliminating nuclear weapons and at the other end the problem of nuclear accidents. Naturally, unforeseeable accidents might not be avoidable, even in a context in which effective safety and security measures against known non-negligible risks are in place and complied with. However, eliminating nuclear weapons and avoiding foreseeable nuclear accidents is presumably possible, at least in principle, notwithstanding the attendant collective action problems.

Collective action problems attendant upon dual use risks exist in the middle of the spectrum and pertain both to R&D that might exacerbate the problem of nuclear weaponry (including 'dirty bombs') and inadequate safety and security measures that might increase or insufficiently reduce the risks of nuclear accidents, sabotage of nuclear facilities or the proliferation of nuclear weapons. Note that while we ought not to confuse these logically distinct collective action problems that exist on our spectrum, it is also important to see how they are connected as is the case, as we saw above, with Iran's nuclear program. Let us begin with the collective action problem of potential nuclear war with a view to identifying a framework for dealing with collective action problems attendant upon dual use risks.

Perhaps the most influential approach to the collective action problem of a potential nuclear war is that of rational choice theory (understood in terms of rational self-interested individual agents engaged in interdependent action), and one of its most controversial solutions that of nuclear deterrence based on the doctrine of mutually assured destruction (MAD). As its name suggests, MAD is unsatisfactory because it does not remove the problem, but at best provides us with a means to live with the problem; nuclear weapons remain but are (supposedly) unlikely to be used on pain of mutual destruction. We say "at best", since the assurance promised by MAD requires that the nuclear 'triggers' are so-called 'hair-triggers' and, as such, inherently risky mechanisms to be relying upon, given the magnitude of the stakes. More generally, for reasons about to be given, rational choice theory is unsuited to this particular kind of collective action problem. Indeed, framing the problem of avoiding the obliteration of the human race, or at least of very large populations of civilians, as a competitive 'game' is entirely perverse.

Individual human agents can, and often do, engage in action-determining reasoning from collective goals and interests to individual actions, including from (but not only from), collective goals and interests to which members of social groups and organisations are strongly morally and/or institutionally committed. Indeed, organisational action, including the actions of members of governments and armed forces, depend on this.

Let us now turn to the analysis of various collective action problems and the role that collective moral responsibility, in particular, might play in their amelioration. As a preliminary to our own analysis we must first identify the deficiencies in the rational choice model standardly used to explicate such problems. The rational choice model assumes rational self-interested individuals in competition with one another. This is fine as far as it goes. However, as Amartya Sen and others have argued, it is far from being the whole of the story.[25] Specifically, it does not leave room for rational individual action performed in the collective self-interest and/or in accordance with socially engendered moral principles and purposes[26]; yet the latter are ubiquitous features of human collective life, including in the economic and political spheres. Importantly, this one-sided fixation with individually rational self-interested action eliminates the possibility, in effect, of finding a solution to collective action problems

---

[25] Sen (2002)

[26] See Elster (1989), 531–52.

of the kind in question. As Elinor Ostrum, the Nobel laureate in economics, quipped[27] in relation to the title of Mancur Olson's famous rational choice monograph, *The Logic of Collective Action*: "It should have been called, *"The Logic of Collective Inaction"*".

What is needed at this point is the acceptance of the above-stated proposition that individual human agents can, and often do, engage in action-determining reasoning from collective goals and interests to individual actions: collective goals and interests to which members of social groups and organisations are strongly committed. On this view of individual reasoning from common goals and collective interests to their own individual action, what is called for is a conception of an individual human agent *qua member of an organisation, nation-state or, indeed, the human race*, e.g. qua scientist, qua human being. In short, the individual internalises the collective goals and interests of the organisation or group to which he or she belongs. Crucially, these collective goals and interests can, and often do, transcend the role occupant's prior and limited, individually rational self-interested, goals and interests; moreover, the collective goals and interests in question can, and often are, embraced by the individuals in question on the grounds that they are desirable from an impartial or, at least, collective standpoint. This capacity of individuals to reason from, and act in accordance with, collective goals and interests is not without its problems. For one thing, the collective goals in question, even if plausibly believed to be in the collective self-interest of the nation-state or other collective in question, might nevertheless be morally problematic from a wider moral perspective (e.g. those of the US or Soviet nuclear weapons program). For another thing, the collective goals and interests in question might be at variance with significant groups within the collective. Thus the collective interest of the members of government might be at variance with the interests of the wider society (e.g. the members of the authoritarian Kim Jong-un regime in North Korea is evidently primarily interested in developing its nuclear weapons program in order to maintain its ascendancy within the North Korean polity and at the expense of the economic and human rights interests of the North Korean citizenry). Or the collective interest of a market-based organisation or sector might be at variance with the interests of the wider society (e.g. the members of a nuclear power company with a business model based on constructing cheaper, but somewhat unsafe, reactors, as is evidently the case with the Russian state-owned company, Rosatom, which is building reactors for (mostly) developing countries).[28]

We suggest that the basic structure of practical means/end reasoning in the case of cooperative schemes, or joint actions more generally, is from mutually believed in shared ends to the performance of contributory individual actions. The shared or collective ends of relevance to our discussion here constitute what are manifestly human goods, such as the avoidance of the (deliberate, culpably negligent or accidental) destruction of the human race or, at least, avoidance of major and perpetual radiological harmful impact on the health of large numbers of people. Moreover, this

---

[27]At a presentation she gave at Delft University of Technology in June 2010 at which Seumas Miller was present.

[28]Taebi and Mayer (forthcoming).

basic structure does not change in the case of complex joint actions, such as those constitutive not only of layered structures of joint actions but also of sub-institutional joint actions, notably joint institutional mechanisms,[29] e.g. NPT.

Viewed from this perspective, generally speaking, the solution to collective action problems is joint action, typically in the form of an enforceable cooperative scheme, in which there is a collective end the realisation of which removes or at least curtails the collective action problem. Notice that for the kinds of collective problems under discussion here, namely, collective action problems in the global nuclear weapons sector, the enforceable cooperative schemes in question are *institutional* arrangements. Moreover, this collective end is a collective good and has motivational force for the participants in the joint action. Naturally, individualist self-interest remains a problem and, as stated, enforcement is an additional requirement that brings its own problems. However, framing the issue in this manner enables us to see: (1) collective self-interest and/or a collective good may in fact be constitutive of individual self-interest, e.g. a nation-state's individual self-interest is constituted in part by its participation in, and contribution to, the realisation of the collective ends of the international community; (2) collective self-interest and/or collective goods may in fact override individual self-interest; individual self-interest does not necessarily dominate collective self-interest and/or collective goods, e.g. the success of the NPT might be more important to a nuclear power company than its own profits; (3) in cases where collective self-interest and/or collective goods have motivational force, but it is overridden by individual self-interest, enforcement is necessary but not sufficient i.e. enforcement mechanisms acting alone are not sufficient for compliance. In relation to (3), we note that given the motivational role of collective self-interest and/or collective goods, enforcement does not have to be sufficient for compliance.

Having provided accounts of collective moral responsibility and collective action problems it is now time to turn to the matter of institutionally embedding collective responsibility in relevant global institutions. This is primarily an exercise in respect of prospective, as opposed to retrospective, moral responsibility. As such, it requires that matters of institutional redesign, implementation and ongoing compliance be attended to.

The generic solution to this kind of collective action problems in question is, at least in part, an enforced cooperative scheme: enforced cooperative schemes are one important way to embed collective moral responsibility in institutional settings suffering from harm inducing collective action problems.

Let us now see how this might work in relation to, firstly, nuclear weapons and, secondly, dual use problems. First, the US and other nations must and, indeed, in large part do, understand that the elimination of nuclear weapons is in everyone's long interest, since it is the only feasible way to ensure against nuclear war and the destruction of the human race (or, at least, a large proportion of the populations

---

[29]Joint institutional mechanisms include such things as voting mechanisms. Such mechanisms involve joint actions but the actual output (e.g. Donald Trump in the case of the US elections) is not something that is being aimed at by all the participants, albeit all bona fide participants are committed to the output of the mechanism whatever that is. See Miller (2016).

of the nuclear powers engaged in nuclear conflicts with one another), given that each country needs to accept to abolish its own weapons for others to do so. Here the avoidance of nuclear war is both the collective end and the collective good in question. Achieving this understanding might be difficult in the case of a nation whose leadership has little or no commitment to the interests of its own people, such as in the case of North Korea. Second, and consistent with their mutual interests (let us assume), the nuclear powers and the US and Russia, in particular, should move to progressively reduce their stockpiles of nuclear weapons, as they have agreed to do in the past in accordance with various treaties such as the Strategic Arms Reduction Treaty (START). Third, and consistent with their mutual interests, the US, Russia and other nuclear weapons and nuclear power states should strengthen and widen the institutional mechanisms already in place, including increasing the membership of NPT, the verification procedures of the IAEA, and the NSG (or some equivalent). Fourth, the US and its allies, in particular, should pressure recalcitrant states to comply with the above steps by means of a range of measures, including sanctions, but also access to new nuclear missile defence systems such as Star Wars.[30] Here there are complications, notably in the case of North Korea. Arguably, China alone has significant leverage to pressure North Korea to comply with the institutional mechanisms in question, albeit even China's leverage appears to be limited and restricted to sanctions. Finally, there is evidently a need for a new international multi-lateral enforcement mechanism,[31] including in respect of related issues such as facilitating the transfer of fissile materials and technology to terrorist groups.

While the prospects for taking all the above five steps seem reasonable, the sticking point might come when all sides, e.g. US, Russia, China and North Korea, must finally eliminate rather than substantially reduce their nuclear weapons programs and stockpiles of nuclear weapons. (We note that unfortunately the process of reducing programs and stockpiles has in fact stalled, if not gone into reversal.) For if one nation preserved its own nuclear program it would have immense political power and have this power even without any intention to annihilate its enemy states, let alone be under threat of annihilation itself. Decades ago, in the context of the Cold War, Reinhold Niebuhr made a remark which evidently remains relevant for us today: "For nuclear disarmament, even if undertaken mutually, involves some risk to the securities of both sides. There is small prospect that either side would be willing to take the risks. This remains true even if their failure to do so would involve the world in the continued peril of nuclear warfare. One may take for granted that neither side actually intends to begin the dread conflict. But it may come upon them nevertheless by miscalculation or misadventure".[32]

What of the collective action problems posed by dual use risks? Here, as we have seen, there are a set of interconnected risks, including R&D for peaceful purposes leading to weaponisation by malevolent secondary users, the possibility of terrorists acquiring 'dirty bombs', and culpable negligence with respect to the risks of nuclear

---

[30]Butler (2001), 152.

[31]Ibid. 154.

[32]Niebuhr (1959), 269.

accidents. We suggested above that the responsibility to engage in dual use harm pre-vention in respect of R&D in the nuclear industry is a collective moral responsibility. We now suggest that it is a collective moral responsibility to design and implement an institutionally-based *web of prevention*. However, this web of prevention should not be specifically focussed on dual use problems per se. Indeed, this web of prevention needs to be designed in a manner such that dual use based-harm is only one source of the overall harm to be prevented. Accordingly, the measures mentioned above in relation to nuclear weapons reduction should be considered to be part of the overall web of prevention. What are some of the other main parts of this web—in so far as the web is to address dual use based-harm?

There is a need for regulatory authorities at national, industry and organisational levels. At the national level there is a need for a raft of safety and security measures, including at nuclear facilities, in relation to transport of nuclear materials with respect to the surveillance and monitoring of terrorists, and so on and so forth.

Amongst other things, international, national, industry, occupational and organisa-tional authorities (or, perhaps in some cases at the organisational and/or occupational levels, advisory committees) should assist in the process of identifying dual use prob-lems in R&D and make adjudications/provide advice, both with respect to actually undertaking the dual use R&D in question and with respect to the dissemination thereof, e.g. in scientific journals.

The regulatory architecture ought to include restrictions on stockpiles and export of nuclear materials, prescribing of the safety and security-based conditions under which R&D can be undertaken and by whom, e.g. background checks and security clearance for research personnel, training programs, licensing of organisations, and so on. Moreover, there is a need for educational and training programs that include material on dual use concerns. Ethics codes and codes of conduct are an important element of such programs. These codes can operate at an organisational and industry-wide level, but also at an occupational level. Thus different occupational groups, e.g. nuclear scientists and engineers might have their own codes, both at a national and international level. Further, elements of the regulatory architecture include protections for professional reporting of misuse of nuclear materials, technology and the like.

More generally, there is a need to restrict or, at least, stem the flow of knowledge in respect of the manufacture of nuclear and radiological material. The knowledge in question is collective (propositional and practical) knowledge and it should, as far as possible, be restricted to relevant responsible expert sub-groups. In short, there is a collective moral responsibility to maintain, where possible, collective public ignorance with respect to the manufacture of nuclear materials and, specifically, curtail transfer of collective expert knowledge of nuclear technology and materials to malevolent actors, including malevolent expert groups. The justification for such knowledge restriction or censorship is well-known. Among other things, it is argued in effect that there is no general moral right to know how to produce WMDs or materials that might potentially cause large-scale harm.

However, it is often responded to this kind of policy prescription that it is too late, that such knowledge is already 'on the internet'. In some cases this might be so, in other cases perhaps not. At any rate, the first general point to be made is that the purpose behind such restrictions is to reduce the risks, not eliminate the risks entirely. For the latter is not possible. The second point is that such restrictions need to work hand-in-glove with other measures and the security task might be less onerous in the context of the restrictions in question. The measures in question include not only regulatory ones, but also the acquisition of new knowledge that might assist by providing the means to protect against the harmful effects of the radioactive material in question.

Specific issues that might need to be addressed include new developments in R&D, and the identification and resolution of collection action problems in otherwise well-regulated nuclear industries. Thus firms in a given industry are in commercial competition. Firm X would strictly comply with the letter and spirit of the regulations concerning dual use R&D if X believed that all other firms did so. What if X does not believe this and that, therefore, X is at a commercial disadvantage? In such a situation in a highly competitive market, firm X (and, by analogy, firms Y, Z etc.) might make dual use cost/benefit judgments that favor commercial self-interest over strict compliance with, say, stringent safety and security regulations. Thus short term organisational commercial interests might override the long term public interest in ensuring dangerous material, for example, are not developed or, if developed, do not fall into the wrong hands.

6.4   Conclusion

In this chapter we have argued that the dual use problem exists in an acute form in the nuclear sciences and technology. For scientific research, technology and materials in the nuclear sciences have enabled, on the one hand, unbounded nuclear energy for peaceful purposes, and yet on the other, massive arsenals of nuclear WMDs with the potential to destroy humankind. The nuclear industry has also facilitated the potential for malevolent actors to deploy 'dirty bombs' and created the conditions under which culpable negligence can result in nuclear disasters. Moreover, scientists and technologists have specific moral responsibilities, both individual and collective, given their central roles in nuclear power plants and government sponsored weapons programs, in particular. These include the collective responsibility to maintain collective public ignorance and not to enable collective expert knowledge in respect of certain expert groups.

The generic solution to the kind of collective action problem to be found in the nuclear sciences, notably the nuclear weapons arms race, is at least in part an enforced cooperative scheme: enforced cooperative schemes are one important way to embed collective moral responsibility in institutional settings suffering from harm inducing collective action problems. This requires widening and strengthening existing institutional arrangements such as the NPT, but also creating additional ones, especially in the area of enforcement.

References

Bunn, Matthew, Martin B. Malin, Nickolas Roth, and William H. Tobey. 2016. *Preventing Nuclear Terrorism: Continuous Improvement or Dangerous Decline?*. Cambridge, MA: Project on Managing the Atom, Belfer Center for Science and International Affairs, Harvard Kennedy School.

Butler, Richard. 2001. *Fatal Choice: Nuclear Weapons and the Illusion of Missile Defense*. 83, 152 and 154. Sydney: Allen and Unwin.

Corera, G., and J.J. Myers. 2009. *Shopping for Bombs: Nuclear Proliferation, Global Insecurity and the Rise and Fall of the A Q Khan Network*. New York: Oxford University Press.

Elster, Jon. 1989. Rationality and Social Norms. In *Logic, Methodology and Philosophy of Science*, 531–52.

Evans, Nicholas G. 2013. Contrasting Dual-use Issues in Biology and Nuclear Science. In *On the Dual Uses of Science and Ethics: Principles, Practices and Prospects*, ed. Brian Rappert, and Michael Selgelid, 262–3. Canberra: ANU Press.

Findlay, T. 2011. *Nuclear Energy and Global Governance: Ensuring Safety, Security and Non-Proliferation*. New York: Routledge.

Findlay, T. 2012. *Unleashing the Nuclear Watchdog: Strengthening and Reform of the IAEA*. Waterloo: Center for International Governance Innovation.

Ferguson, Charles D. 2011. *Nuclear Energy: What Everyone Needs to Know*. New York: Oxford University Press.

Goldberg, S.M., and R. Rosner. 2011. *Nuclear Reactors: Generation to Generation*. Cambridge, MA: American Academy of Arts and Sciences.

Meier, Oliver, and Iris Hunger. 2014. *Between Control and Cooperation: Dual-use, Technology Transfers and the Non-Proliferation of Weapons of Mass Destruction, 17*. Osnabruck: DSF.

Miller, Seumas. 2013. Collective Responsibility, Epistemic Action and the Dual Use Problem in Science and Technology. In *On the Dual Uses of Science and Ethics*, ed. Brian Rappert, and Michael Selgelid. Canberra: ANU Press.

Miller, Seumas. 2016. Joint Action: Some Applications. *Journal of Applied Philosophy*.

Miller, Seumas. 2017. Ignorance, Technology and Collective Responsibility. In *Perspectives on Ignorance from Moral and Social Philosophy*, ed. Rik Peels, 217–237. Oxford: Routledge.

Miller, Seumas, and Michael Selgelid. 2007. Ethical and Philosophical Consideration of the Dual Use Dilemma in the Biological Sciences. *Science and Engineering Ethics* 13: 523–580.

Miller, Steven E. and Scott D. Sagan. 2009. Nuclear Power without Nuclear Proliferation? *Daedalus*, Fall: 11.

Morland, Howard. 2005. Born Secret. *Cardozo Law Review* 26(4): 1401-8 and 1402.

Niebuhr, Reinhold. 1959. *The Structure of Nations and Empires, 269*. New York: Charles Scribner.

Rubin, A. and M. Schreurer. 2016. Belgium Fears Nuclear Plants are Vulnerable. *International New York Times*, March 25. http://nyti.ms/22LKA1P.

Schweber, Silvan S. 2000. *In the Shadow of the Bomb: Bethe, Oppenheimer and the Moral Responsibilities of the Scientist*. Princeton: Princeton University Press.

Scott, L., and R.G. Hughes (eds.). 2015. *The Cuban Missile Crisis: A Critical Reappraisal*. London: Routledge.

Sen, Amartya. 2002. *Rationality and Freedom*. Cambridge, Mass.: Harvard University Press.

Taebi, Behnam. 2012. Intergenerational Risks of Nuclear Energy. In *Handbook of Risk Theory. Epistemology, Decision Theory, Ethics and Social Implications of Risk*, ed. S. Roeser, R. Hillerbrand, P. Sandin and M. Peterson, 295–318 (303–304). Dordrecht: Springer.

Taebi, Behnam. 2013. Moral Dilemmas of Uranium and Thorium Fuel Cycles. In *Social and Ethical Aspects of Radiation Risk Management* ed. D. Oughton and S.O. Hansson, 259–280. Amsterdam: Elsevier.

Taebi, Behnam, and J.L. Kloosterman. 2015. Design for Values in Nuclear Technology. In *Handbook of Ethics, Values, and Technological Design: Sources, Theory, Values and Application Domains*, ed. J. van den Hoven, P. Vermaas, and I. Van de Poel, 805–829. Dordrecht: Springer Science+Business Media.

Taebi, Behnam and M. Mayer. Forthcoming. By Accident or by Design: Pushing the Global Governance of Nuclear Safety. *Progress in Nuclear Energy*.

Taebi, Behnam and S. Roeser. 2015. The Ethics of Nuclear Energy. An Introduction. In *The Ethics of Nuclear Energy. Risk, Justice and Democracy in the post-Fukushima Era*, ed. B. Taebi and S. Roeser, 1–14 and 2. Cambridge: Cambridge University Press.

Chapter 7
Cyber-Technology

Abstract Cyber-technology is a new and emerging area of dual use concern. Consider autonomous robots. On the one hand, autonomous robots can provide great benefits, e.g. providing for the health and safety of elderly invalids. On the other hand, autonomous robots have the potential to enable great harm, e.g. weaponised autonomous robots (so-called 'killer robots'). As we have seen, the intended great harm is typically delivered by a weapons system of some sort, e.g. chemical, nuclear or biological weapons. Cyber-technology is apparently no different in this respect since, after all, there are so-called cyber-weapons, such as the Stuxnet virus used to shut down Iranian nuclear facilities. In this chapter the definition of dual use technology elaborated in Chap. 2 is modified in light of some distinctive properties of cyber-technology. This modified definition is applied to cyber-technology with a view to identifying cyber-technologies that are dual use technologies. It is concluded that weaponised autonomous robots, various forms of computer viruses, and ransomware are dual use technologies, but that the internet and other forms of cyber-infrastructure are not.

Having considered dual use issues in the established fields of the chemical and nuclear sciences it is now time to turn to the new and emerging area if cyber-technology. Thus, on the one hand, autonomous robots might provide great benefits, e.g. providing for the health and safety of elderly invalids. On the other hand, autonomous robots have the potential to enable great harm, e.g. weaponised autonomous robots (so-called 'killer robots').

This chapter was co-authored by Seumas Miller and Terry Bossomaier.

© The Author(s) 2018   91
S. Miller, *Dual Use Science and Technology, Ethics and Weapons of Mass Destruction*, SpringerBriefs in Ethics, https://doi.org/10.1007/978-3-319-92606-3_7

As we have seen, the intended great harm is typically delivered by a weapons system of some sort, e.g. chemical, nuclear or biological weapons. Cyber-technology is apparently no different in this respect since, after all, there are so-called cyber-weapons, such as the Stuxnet virus used to shut down Iranian nuclear facilities.[1] Moreover, as we have seen in earlier chapters, the intended harm might be caused by something other than a weapons system. For instance, a homicidal lunatic might dump an extremely dangerous man-made toxin into a city's supply of clean water with the intention of killing a large number of the city's residents. In this situation the R&D that enable the production of the toxin might well be regarded as dual use in character. However, the toxin is not per se a weapon.

Further the harm in question is not merely epistemic harm in the sense of harm consisting merely of believing what is false or of being in a state of ignorance. Of course, epistemic harm may lead to non-epistemic harm. For instance, ignorance of the toxic nature of some liquid may result in a child or even the members of a whole community drinking it and suffering death as a consequence. But that is another matter; for death is not in and of itself an epistemic harm.

In this chapter we apply the definition of dual use technology elaborated in Chap. 2 and apply it to cyber-technology with a view to determining which types of cyber-technology, if any, are dual use in character. In doing so we modify the existing definition somewhat. Importantly, according to this (modified) definition, cyber-technology used to effect mass destruction and in which the weapons used are controlled by computers, including with respect to the selection of targets (and, perhaps the selection of the weapons themselves), constitutes dual use technology, as do various forms of computer viruses and ransomware. Given this focus on the definitional issues raised by dual use cyber-technology, space does not permit us to address regulatory matters in this chapter in the manner done so in the chapters on chemical (Chap. 5), nuclear (Chap. 6) and biotechnology (Chap. 8). So these regulatory matters in relation to cyber-technology, important as they are, will need to be left for another occasion.

7.1   Epistemic Character of Cyber-Technology

Before turning to our main task of applying this definition of dual use to cyber-technology, we need to address an important problem posed by cyber-technology for the definition as it stands, namely, the epistemic character of cyber-technology. Thus in a plane with a nuclear bomb it is the bomb which is the weapon. If the plane is controlled by a robot (in whatever sense of control robots control things) which can decide on targets, is the robot dual use? After all, it is not actually the destructive agent.

Now consider the case of a truck as a lethal weapon, for example, a petrol tanker, which is deliberately driven into a building, causing a massive explosion with many

[1] Kelley (2013), Karnouskos (2011), 4490–4494 and Shearer (2010).

casualties. The relevant elements of this scenario are: the driver; the truck; the road; the petrol. From the account of dual use proffered in Chap. 2 it seems that the only candidate for being dual-use, specifically dual use material rather than dual use technology, is the petrol; the truck is not dual use technology, since in and of itself it is analogous to the baseball bat used repeatedly as a weapon (see Chap. 2, Sect. 2.1) or a knife used in a series of attacks against a large number of victims. Indeed, in the recent terrorist attack in Nice in France on Bastille Day in 2016 a truck was used as a weapon and killed dozens of innocent persons.[2] However, this was hardly on the scale of a WMD attack and, in any case, arguably consisted of a series of attacks as the truck hit one person after another. This raises a problem in relation to putative instances of dual use cyber-technology, since it is the things cyber controls which cause the damage rather than the computer system itself.

Compare the petrol truck example to an improvised explosive device (IED) detonated by a terrorist using a mobile phone. The elements are: the terrorist (driver); the mobile phone (truck); the mobile carrier network (road); and the explosive (petrol). Is the explosive used in this terrorist attack the only candidate for being dual use? In short, is the mobile phone IED example precisely analogous to the petrol truck example?

One important difference is that the truck could have caused the explosion without being driven on the road; for instance, the truck could have been driven, let us assume, on the grass area adjacent to the road. Roads are not an essentially enabling infrastructure for trucks. By contrast, the mobile phone could not have been used to trigger the IED without using a mobile phone network and the internet. The latter are essentially enabling technological infrastructure for using the mobile phone to trigger the IED.

Now consider a variant of the Stuxnet virus which causes a nuclear power station meltdown. The elements are: programmers who build and launch the virus (driver); the internet (road); the virus (truck); and the power station or its fissile material (explosive). There are two relevant points of difference between this kind of case and our truck driver example. The first point of difference is the one already mentioned in relation to the mobile phone IED. The Stuxnet virus, unlike the truck, relies on an essentially enabling technological infrastructure, namely, the internet.

However, there is a second relevant point of difference. Moreover, this also differentiates the Stuxnet virus from the mobile phone IED, namely, the so-called 'autonomy' of the Stuxnet virus. The virus is in some sense or to some degree outside the control of the humans who have unleashed it and, in addition, the virus in some sense or to some degree exercises control over some of the objects it interacts with, e.g. the computers it infects, machinery 'controlled' by computers (In this respect it is akin to a biological virus created and released by some malevolent scientist.). Let us refer to this kind of autonomy or control as computer autonomy or computer control to distinguish it from human autonomy/control. The notions of human autonomy, human freedom, and human control are inherently difficult and essentially contested. For instance, many argue that human autonomy implies moral

---

[2]BBC News (2016).

agency, the ability to choose ultimate ends, and so on. The point to be made here is that it is by no means clear that computers could possess these properties (We return to this issue in Sect. 7.2 below.).

We are now in a position to utilize the notion of essentially enabling technology and computer autonomy to distinguish two cases of truck terrorism:

1. A terrorist hacks into a truck's computer system and diverts it into a crowd of people. The internet and computer interaction with machinery are essentially enabling technologies for the *terrorist hacker's* remote control of the weapon (the truck).
2. The terrorist writes a computer worm (see Sect. 7.3 below), which hunts for trucks on the Internet of Things (IoT), and causes those it infects to drive into a crowd of people. The internet and computer interaction with machinery are essentially enabling technologies for the *computer worm's* control of the weapons (the trucks).

The second example involves computer autonomy (as opposed to human autonomy) in the selection of trucks and targets and, therefore, the cyber-technology is conceptually integral to weapons of destruction in a manner in which the essential enabling technology of the internet, let alone roads, is not. In our terrorist hacker example the weapon is the truck and it is selected by, and under the control of, the human hacker (albeit the terrorist hacker only controls the weapon remotely and indirectly and in doing so relies on the essentially enabling technology of the computer interaction with machinery and the internet). By contrast, in our computer worm example, while the weapons are trucks, they are selected by, and under the 'control' of, the computer worm: the cyber technology. Accordingly, the cyber-technology consisting of the computer worm utilizing the essentially enabling technology of the internet and computer interaction with machinery is conceptually integral in a strong sense to the weapons of destruction (the trucks).

The upshot of this discussion is that cyber-technology while epistemic in character can, nevertheless, be conceptually integral in a strong sense to weaponry. Arguably, therefore, the epistemic character of cyber-technology does not necessarily prevent it from being dual-use technology. Moreover, as is illustrated by our computer worm example, cyber-technology can be used to kill very large numbers of people following on the release of a single virus. We conclude that cyber-technology, such as computer worms, used to effect mass destruction may well constitute dual use technology. Indeed, in Sect. 7.3 below we argue that computer worms (and related computer viruses), autonomous robots and encryption-utilizing ransomware, in particular, are in fact species of dual use cyber-technology, at least in certain configurations. However, before arguing that these *are* instances of dual use cyber-technology, we need to argue for the proposition that the internet and certain other cyber-technologies are *not* species of dual use technology, notwithstanding the tendency to believe that they are. Critical cyber-technological infrastructure such as the internet, for instance, is often referred to as dual use.

7.2 Identifying Dual Use Cyber-Technology

Recent technological developments in information and computer technology and in artificial intelligence have given rise to dual use problems. However, here we need to be circumspect. We have defined dual use science and technology as having both beneficial and harmful purposes—where the harmful purposes typically, albeit not necessarily, involve weapons and, paradigmatically, weapons of mass destruction.

Infrastructure, such as dams, telephone cables and power-lines, if deliberately destroyed or severely damaged for a prolonged period by weapons in the context of war, may lead to widespread suffering, even death. However, it would not follow that such critical infrastructure was dual use in our sense. Of course, such infrastructure may well be dual use in the quite different sense that it is used by both civilians and the military. Moreover, its destruction may harm both civilians and the military. So the population at large is vulnerable to great harm by virtue of its dependence on critical infrastructure. However, the infrastructure in and of itself is not a weapon or other vehicle being used to harm; rather it is the thing being damaged or destroyed (from which harm to the population results).

The internet is critical infrastructure; indeed, critical global infrastructure (and, indeed, as we saw above, potentially essential enabling technology for weapons of mass destruction). A good deal of interpersonal, organisational, local, national, international etc. communication and data transfer is now dependent on the internet. Accordingly, central national and global institutions are dependent on the internet. For example, the global financial system depends on the internet. However, this dependence makes these institutions and, therefore, the societies in part constituted by these institutions extraordinarily vulnerable should this critical infrastructure, or important parts of it, be severely damaged for a prolonged period by, say, terrorists. Moreover, the internet is used by civilians and military alike. So the population at large, indeed multiple populations, are vulnerable to great harm by virtue of their dependence on the internet. Nevertheless, for the reasons given above in respect of other types of critical infrastructure, the internet per se is not dual use technology in our sense.

Developments in communication and information technology (ICT) not only enable the provision of critical infrastructure, they also enable the efficient collection, storage, analysis, communication and dissemination of information on an unprecedented scale. Consider, for example, social media, such as Facebook or Twitter. Consider also Big Data.[3] Big Data simply means all the data in some domain; for example, all the financial transactions in a global capital market in a 24 h period.

Facebook and Twitter enable the immediate communication of information to vast audiences and this has had a revolutionary effect on, for instance, political campaigns in the USA, such as that of Barack Obama, Hilary Clinton and Donald Trump. Again, the collection, storage and analysis of Big Data is an extraordinary treasure-trove for those seeking to benefit humankind, e.g. for demographers projecting future population numbers or climate scientists trying to determine the rate of global warming.

---

[3] Mayer-Schonberger and Cukier (2013).

Of course, social media and Big Data are also able to be used for harmful purposes. Terrorists use social media to recruit, incite and provide access to training manuals, e.g. how to make an IED. Authoritarian governments use Big Data to monitor their citizens intrusively and, thereby, violate their civil liberties.

Nevertheless, neither social media nor Big Data are dual use technologies in our sense. For the ultimate weapon-based harm done by terrorists who use social media, namely, the murdering of innocent people is not directly done by the essentially communicative acts performed by terrorists on social media. Social media is not per se a weapon as, for example, is a nuclear warhead; nor is social media weaponised as, for example, is an aerolised pathogen in a container fitted to a weapons delivery system. Again the ultimate weapons-based harm done by authoritarian governments who collect and analyse data about their citizens, namely, the forcible incarceration, torturing and/or murdering of their citizens, is not directly done by the essentially epistemic acts performed by those who collect and analyse this data. Naturally, the collection and analysis of some of this data, e.g. personal information of citizens, may constitute a violation of the privacy rights of the citizenry and may, as such, be morally wrong. But dual use technology, as we are using the term, typically involves weapons-based harm and, in any case, harm of a more serious kind than mere violation of individual privacy rights. A camera, for example, is not dual use technology in this sense merely because it could be used to violate someone's privacy rights.

For a similar reason technology that enables cyber-theft is not as such dual use technology in our sense. Theft does not necessarily involve weapons-based harm; so it does not meet this important criterion (Although, as we have argued throughout this work, dual use harm is not necessarily weapons-based.). Moreover, theft of property is, other things being equal,[4] at the lower end on the scale of harms. A screwdriver, for example, is not dual use technology merely because it could be used to open a locked box and enable the theft of the contents. Of course, in the case of cyber-theft the 'item' stolen is typically intellectual property, for example data, algorithms. Being theft of intellectual property, cyber-theft does not necessarily deprive the owner of the use of the property, although the owner may well be deprived of many of the rights and benefits of ownership, such as exclusive use and the economic benefits that flow from exclusive access.

Cyber-theft needs to be distinguished from cyber-espionage. The latter refers to the theft by some computer-based means (as opposed to, for example, by physical removal of paper-based documents): (i) of data or other intellectual property stored in an ICT system; (ii) that is reasonably regarded as confidential from a national security perspective; (iii) in order to realize some political or military purpose. Here the Snowden case is salient.[5] Edward Snowden was a low level private contractor to the NSA who breached legal and moral confidentiality obligations by engaging in

---

[4]Naturally, other things might not be equal. Theft of a person's means of livelihood may put their life at risk.

[5]Harding (2014).

unauthorized accessing, retrieving and/or releasing of a large volume of confidential data from the NSA to the international press.

Investigations by the US computer-security firm, Mandiant, indicate that China is a major cyber-thief.[6] For there are multiple acts of cyber-theft originating from the headquarters of the China's People's Liberation Army Unit 61398. Indeed, according to Mandiant most cyber-attacks on US corporations, US infrastructure (e.g. power grids) and US government agencies originate from China and China's large scale cyber-theft comprises hundreds of terabytes from 140 countries.

Notwithstanding the ultimate harm done by cyber-theft or cyber-espionage, technology used to perform such actions is not dual use technology in our favoured sense. For cyber-theft and cyber-espionage are acts of theft, specifically, theft of intellectual property. However, the possession by another person of one's intellectual property is essentially an epistemic condition and, as such, does not constitute a serious harm to oneself. Rather it is what the person can do as a result of his or her new found knowledge that is potentially profoundly harmful. Accordingly, the fact that cyber-technology is vulnerable to acts of cyber-theft and cyber-espionage does not make it dual use technology.

Thus far in this section we have identified various harmful uses of certain forms of cyber-technology and argued that, nevertheless, the technology in question is not dual use technology in our sense. The time has now come to discuss those species of cyber-technology that are dual use technology. We consider two such species, namely (i) computer viruses and (ii) autonomous robots, and a third putative species, (iii) encryption or, at least a use thereof, ransomware.

7.3 Dual Use Cyber-Technology: Viruses, Autonomous Robots and Encryption

7.3.1 Computer Viruses

Computer viruses are akin to pathogens. They are potentially extraordinarily destructive weapons; indeed, they are potentially WMDs. However, like their biological counterparts, computer viruses are not necessarily harmful, nor do they necessarily hide themselves. They are essentially self-replicating programs which install themselves in computers without necessarily having the consent of the computer user. Moreover, the software technology underpinning computer viruses is extraordinarily beneficial. It is essentially the technology that enables the construction of software agents that can collect, transmit, encrypt etc. information. Accordingly, this technology is dual use technology.

For the last half-century computers emulated the running of many programs simultaneously, through the mechanism of time-sharing. Most of these programs were

---

[6]Mandiant Intelligence Centre (2016).

neither started by, nor communicate with, any human users. They do things like manage the file system, control network traffic and other housekeeping things. But in the last two or three decades another type of autonomous program has appeared, the computer virus, an example of computer malware. It may arrive in several ways, either via computer networks or files copied from portable media.

The first worm to gain significant notoriety was the eponymous Morris Worm, which got its inventor, a bright and probably well intentioned postgraduate student (now with tenure at MIT), into serious trouble. The idea of a worm is that it is a standalone program, which replicates itself and finds routes to installing itself on other computers. Morris misjudged the replication rate, and, multiply infected computers were brought to a standstill, possibly as many as 10% of the computers on the network. Clifford Stoll estimated the economic damage of up to $10 million.[7]

We can classify malware for our present discussion into three categories: local; device-oriented; and global. Local malware does things which will usual impact a single, or small group of users, such as encrypting the hard disc, so-called ransomware. We consider this below in Sect. 7.3.3 so we won't consider it further here. "Device-oriented" means that the malware is out to attack a controller of some physical device. If the device has the potential to cause widespread destruction, then this would fit our dual-use definition. The destruction does not have to be human, at least directly. It could be costly infrastructure, machinery, or even a virtual entity such as a stock market. The most remarkable such piece of malware in recent years was the Stuxnet virus mentioned above. A computer worm, it trundled around the internet until it found the Iran uranium purification centrifuges. It then proceeded to damage them irreparably by increasing their rotation speeds beyond safe limits. The origin of the code is unknown, but the US/Israel are the chief suspects. However, the code itself has been made publicly available and can be readily downloaded. The Stuxnet virus was an example of a worm, which was trying to find something.

The global category of malware is intent on getting to as many machines as possible, sometimes with a specific singular goal in mind. One such common occurrence is a Distributed Denial of Service Attack (DDoS). The Morris worm brought computers down by soaking up compute cycles with more and more copies of the worm. A DDoS brings machines to a standstill by flooding them with internet data packets. However, a single machine is not usually sufficiently powerful to send enough packets. Hence a lot of machines are conscripted in a so-called *botnet*. In this case a computer worm has infected a network of computers (the botnet), with a piece of malware which will flood the target machine with packets from every node in the botnet. The Mirai botnet malware was used in a number of sensational attacks, notably on Dyn, a provider of DNS (Domain Name System) servers, halting GitHub, Twitter, Reddit, Netflix, AirBnb and others. Mirai achieved its attack through using not desktop computers or laptops, but computers on the Internet of Things, cameras, fridges and other smart (or not-so-smart) devices. Just like Stuxnet, the software is now readily available.

---

[7]Spafford (1989), Eisenberg et al. (1989).

Other recent high profile cyber-attacks include the following[8]: the DDoS on Estonian banks,[9] media and government web sites in 2007 perpetrated (it is presumed) by Russia; the above-mentioned Stuxnet malware attack—in which the software worm, Stuxnet, was used to disrupt Iran's nuclear enrichment ICT (information and communication technology) infrastructure in the context of a joint US and Israeli operation (Olympic Games) established to disrupt Iran's nuclear program[10]; Operation Orchard—the Israeli bombing of a Syrian nuclear facility after they had penetrated Syrian computer networks and 'turned off' Syrian air defence systems.

In the case of the DDoS on Estonia there were no deaths or destruction of property and computer technicians unblocked the networks relatively quickly thereby ensuring the disruption was minimal.[11] By contrast, Operation Orchard involved the Israeli bombing of a Syrian nuclear facility immediately after an Israeli cyber-attack on Syrian air defence systems.[12] Stuxnet while targeted at Iranian ICT infrastructure also caused collateral damage by contagion; it infected and shut down computers and computer networks in places such as Indonesia and India.[13]

In Sect. 7.1 we argued that computer worms and, by implication, many other forms of computer virus were potentially dual use technologies, notwithstanding their epistemic character, since they can be conceptually integral in a strong sense to weaponry. This cleared the way for computer viruses to be considered as a candidate for dual use technology. In this section we have provided the evidence that this cyber-technology has been used as a weapon, indeed a weapon of war, and could easily be used as a WMD. We conclude that computer viruses are a species of dual use technology.

7.3.2 Autonomous Robots

Autonomous robots are able to perform many tasks for more efficiently than humans, e.g. tasks performed in factory assembly lines, auto-pilots, driver-less cars; moreover, they can perform tasks dangerous for humans to perform, e.g. defuse bombs. However, as we saw above autonomous robots can also be weaponised and can great harm. Are weaponised autonomous robots a species of dual use technology?[14]

Science fiction movies, such as the Terminator series, have accustomed us to images of armed computerized robots led by leader robots fighting wars against human combatants and their human leaders. The reality is somewhat different. It essentially consists of new and emerging (so-called) autonomous robotic weaponry.

---

[8] See Singer and Friedman (2013). On the Stuxnet and Estonia cases, see also Rid (2013), 32–34.
[9] Lesk (2007).
[10] Sanger (2013).
[11] Rid (2013), Finn (2007).
[12] Rid (2013), 42–43; BBC News (2007), Follath and Stark (2009).
[13] Anwer (2012), Bachrach (2013).
[14] An earlier version of the material in this section appeared in Miller (2015), 153–166.

Consider, for example, the Samsung stationary robot which functions as a sentry in the demilitarized zone between North and South Korea.[15] Once programmed and activated, it has the capability to track, identify and fire its machine guns at human targets without the further intervention of a human operator. Predator drones are used in Afghanistan and the tribal areas of Pakistan to kill suspected terrorists. While the ones currently in use are not autonomous weapons they could be given this capability in which case, once programmed and activated, they could track, identify and destroy human and other targets without the further intervention of a human operator. Moreover, more advanced autonomous weapons systems, including robotic ones, are in the pipeline.

So autonomous weapons are a weapons system which, once programed and activated by a human operator, can—and, if used, do in fact—identify, track and deliver lethal force without further intervention by a human operator. By 'programmed' we mean, at least, that the individual target or type of target has been selected and programmed into the weapons system. By 'activated' we mean, at least, that the process culminating in the already programmed weapon delivering lethal force has been initiated. This weaponry includes weapons used in non-targeted killing, such as autonomous anti-aircraft weapons systems used against multiple attacking aircraft or, more futuristically, against swarm technology (for example multiple lethal miniature attack drones operating as a swarm so as to inhibit effective defensive measures); and ones used or, at least, capable of being used in targeted killing (for example a predator drone with enhanced face-recognition technology such that there is no need for a human operator to confirm a match).

We need to distinguish between so-called 'human in-the-loop', 'human on-the-loop' and 'human out-of-the-loop' weaponry. In the case of human-in-the-loop weapons the final delivery of lethal force (for example by a predator drone), cannot be done without the decision to do so by the human operator. In the case of human on-the-loop weapons, the final delivery of lethal force can be done without the decision to do so by the human operator; however, the human operator can override the weapon system's triggering mechanism. In the case of human out-of-the-loop weapons, the human operator cannot override the weapon system's triggering mechanism; so once the weapon system is programmed and activated there is, and cannot be, any further human intervention.

The lethal use of a human-in-the-loop weapon is a standard case of killing by a human combatant and, as such, is presumably, at least in principle, morally permissible. Moreover, other things being equal, the combatant is morally responsible for the killing. The lethal use of a human-on-the-loop weapon is also akin to long-standing kinds of weaponry (e.g. automatic weapons) and, as such, is presumably, at least in principle, morally permissible. Moreover, the human operator is, perhaps jointly with others, morally responsible, at least in principle, for the use of lethal force and its foreseeable consequences. So it is really only the human out of the loop weaponry that should be regarded as morally problematic. Indeed, some have claimed that no-one is morally responsible for killings done by human out-of-the-loop

---

[15] Shor (1999).

weapons—so-called 'killer-robots'.[16] This so-called responsibility gap is doubtful. Consider the case of a human in-the-loop or human-on-the-loop weapon. Assume that the programmer/activator of the weapon and the operator of the weapon at the point of delivery are two different human agents. If so, then other things being equal they are jointly (that is, collectively) morally responsible for the killing done by the weapon.[17] No-one thinks the weapon is morally or other than causally responsible for the killing. Now assume this weapon is converted to a human out-of-the-loop weapon by the human programmer-activator. Surely this human programmer-activator now has full individual moral responsibility for the killing. To be sure there is no human intervention in the causal process after programming-activation. But the weapon has not been magically transformed from an entity only with causal responsibility to one which now has moral or other than causal responsibility for the killing.

Where does this leave us with our question as to whether weaponised autonomous robots are dual use technology? Moreover, what is the relevance to this question of the distinctions between in the loop, on the loop and out of the loop?

Weaponised autonomous robots are, we suggest, a species of dual use technology, irrespective of whether they are in the loop, on the loop or out of the loop. Here there are a number of considerations (see Sect. 7.1). Firstly, once weaponised, autonomous robots are conceptually integral in a strong sense to their weapons; that is, they utilize the essentially enabling technology of the internet and computer interaction with machinery (the weapon). Secondly, autonomous weapons have the potential to be armed with WMDs, e.g. chemical or nuclear devices.

However, human out-of-the-loop autonomous weapons have a degree of *computer* autonomy that the human in the loop or on the loop autonomous weapons do not. Nevertheless, it simply does not follow from this that the humans who designed, implemented and used out of the loop autonomous weapons are not morally responsible for the killings done by these weapons; specifically, they are collectively i.e. jointly, morally responsible (see Chap. 4). In short, in the case of autonomous weapons, computer autonomy underpins (in part) the conceptual integration of the cyber-technology with the weapon—and, thereby, justifies the claim that autonomous robots are a species of dual use technology. However, here, as elsewhere, the autonomy in question (computer autonomy) should not be confused with human autonomy or be taken to have extinguished human moral responsibility.

7.3.3 Encryption and Ransomware

Some terminological clarifications are useful to begin with. Encryption and decryption are usually paired and a part of the general notion of cryptography, Cryptography goes back to ancient times, maybe even to stone tablets, and is thus neither a digital

---

[16]See Sparrow (2007), 63–77. For criticisms see Steinhoff (2013).

[17]Moreover, each is fully morally responsible; not all cases of collective moral responsibility involve a distribution of the quantum (so to speak) of responsibility.

of silicon-based computer technology. Encryption may be taken to be an algorithm in the first instance, along with its decryption dual. As such it is also neither digital nor silicon. An algorithm, such as RSA,[18]can be carried out with pen and pencil, while the Shor algorithm to break it cannot be carried out at full speed on *any* computer currently available (it requires quantum computation) Thus our concern is with encryption used digitally in silicon and we loosely refer to both encryption and decryption loosely as encryption. However, a novel and, for its time, powerful technology was the Enigma machine used purely for decryption.

We can distinguish three types of encryption which overlap to some extent:

1. encryption of documents—articles, memos, emails, text messages, anything involving natural text (the Enigma Machine falls into this category);
2. encryption for access—passwords, biometrics;
3. encryption for control—machines, robots, weapons.

Since encryption is about transforming information, it cannot physically do any harm. Thus it apparently fails to meet the definition of dual-use. Let us, however, consider the matter further.

Encryption offers enormous benefits. The whole of e-commerce depends upon being able to feed credit card numbers safely into a web site, relying on the https encryption protocol. Encryption of course appears throughout the ages in a military context. Turing's Enigma Machine[19] saved many lives in the Second World War. Yet the encryption itself, or the breaking thereof, was one step removed from harm, which in this case was done by U-boats, or the torpedoes they launched.

However, it is possible to use encryption as a weapon. Consider the victim of ransomware, where the data on a hard disc is maliciously encrypted for financial or other gain. In a ransomware attack, the target computer becomes infected by a piece of so-called malware. It may arrive from an email attachment, a dubious website, a Trojan horse app downloaded or by other means. Once installed it then 'autonomously' sets about encrypting the hard disc, rendering unstable to its owner. Decryption requires a key, for which a ransom is required.

But as we have alluded earlier, dual-use requires harm to a significant number of people. Consider the case of a large hospital. Patient records, treatment procedures and schedules are now kept online. A terrorist hacker could gain access to this system and encrypt the contents of patient records, or perhaps encrypt records of just selected unspecified patients. Many hundreds of patients could suffer serious harm, including death, as a result.

Deleting large amounts of data would not be so effective, since it would usually be possible to restore the data from backups. But a similar effect could be achieved by maliciously altering selected records, changing drug dosage for example. Such mechanisms may be released on the internet to attack autonomously and indiscriminately.

Lest it be thought that ransomware is a minor issue consider the following. Ransomware is thought to cost the Australian economy $1 billion per year, but it usually

---

[18]Rivest et al. (1978), 120–126.
[19]Turing (1939).

operates on a single machine at a time. A user opens an email attachment, which contains the ransomware, which then encrypts their computer. However, in the recent case of WannaCry it was attached to a worm, which spread very rapidly, dropping the ransomware on each computer it affected. Many organisations were compromised, including Deutsche Bahn (German railway), Telefonica (Spanish telecom) and the National Health Service in the UK. The latter case raises the policy issue in a big way.[20] The NHS has been under financial pressure for some time, despite its eulogy in the London Olympics Opening Ceremony. The cost of upgrading software systems was one which could be delayed. There is a further complication, in that when an operating system upgrade is a major one, specialised software may not work properly, if at all. Hence it is not just the cost of the operating system upgrade, but of all the testing and updating of other software. In some cases such legacy software may be very old and extremely difficult to update with confidence that new bugs will not be introduced.

7.4 Conclusion

In this chapter we have taken as our starting point the definition of dual use technology elaborated in Chap. 2 and modified it in light of some distinctive properties of cyber-technology. We have then applied this modified definition to cyber-technology with a view to identifying cyber-technologies that are dual use technologies. We have concluded that weaponised autonomous robots, various forms of computer viruses, and ransomware are dual use technologies, but that the internet and other forms of cyber-infrastructure are not.

References

Anwer, Jared. India Caught in Crossfire of Global Cyber War. *Times of India*, August 20, 2012. http://timesofindia.indiatimes.com/tech/it-services/India-caught-in-crossfire-of-global-cyber-war/articleshow/15567180.cms. Accessed 24 June 2014.
Bachrach, Judy. 2013. The Stuxnet Worm Turns. *World Affairs Journal*, January 30, 2013. http://www.worldaffairsjournal.org/blog/judy-bachrach/stuxnet-worm-turns. Accessed 24 June 2014.
BBC News. 2007. Estonia Hit by 'Moscow Cyber War. *BBC News*, May 17, 2007. http://news.bbc.co.uk/2/hi/europe/6665145.stm. Accessed June 24, 2014.
BBC News. 2016. Nice Attack: What We Know About the Bastille Day Killings. *BBC News*, August 18, 2016. www.bbc.com/news/world-europe-36801671.
Eisenberg, T. et al. 1989. The Cornell commission: On Morris and the Worm. *Communications of the ACM* 32(6): 706–709.http://portal.acm.org/citation.cfm?doid=63526.63530.
Finn, Peter. 2007. Cyber Assaults on Estonia Typify a New Battle Tactic. *Washington Post*, May 19, 2007. http://www.washingtonpost.com/wp-dyn/content/article/2007/05/18/AR2007051802122.html. Accessed June 24 2014.

---

[20]Mattei (2017).

Follath, Erich and Holger Stark. 2009. The Story of 'Operation Orchard': How Israel Destroyed Syria's Al Kibar Nuclear Reactor. *Spiegel Online International*, November 2, 2009. http://www.spiegel.de/international/world/the-story-of-operation-orchard-how-israel-destroyed-syria-s-al-kibar-nuclear-reactor-a-658663.html. Accessed June 24 2014.

Harding, Luke. 2014. *The Snowden Files: The Inside Story of the World's Most Wanted Man*. London: Guardian Books.

Karnouskos, Stamatis. 2011. Stuxnet Worm Impact on Industrial Cyber-physical System Security. In *IECON 2011-37th Annual Conference on IEEE Industrial Electronics Society*, 4490–4494. London: IEEE.

Kelley, Michael. 2013. The Stuxnet Attack On Iran's Nuclear Plant Was 'Far More Dangerous' Than Previously Thought. *Business Insider*. http://www.businessinsider.com/stuxnet-was-far-more-dangerous-than-previous-thought-2013–11?international=true&r=US&IR=T.

Lesk, Michael. 2007. The New Front Line: Estonia Under Cyberassault. *IEEE Security & Privacy* 5(4) (July–Aug).

Mandiant Intelligence Centre. 2016. APT1: Exposing One of China's Cyberespionage Units, Mandiant Intelligence Centre. http://184.72.240.198/.

Mattei, Tobias A. 2017. Privacy, Confidentiality and Security of Health Care Information: Lessons from the Recent Wannacry Cyberattack. *World Neurosurgery* 104: 972–974.

Mayer-Schonberger, V., and K. Cukier. 2013. *Big Data: A Revolution that Will Transform How We Live, Work and Think*. London: John Murray.

Miller, Seumas. 2015. Collective Responsibility for Robopocolypse. In *Super Soldiers: The Ethical, Legal and Social Implications*, eds. Jai Galliott and M. Lotze, 153–166. Aldershot, U.K.: Ashgate.

Rid, Thomas. 2013. *Cyber War Will Not Take Place, 30–32, 32–34 and 42–44*. New York: Oxford University Press.

Rivest, R.L., A. Shamir and L. Adleman. 1978. *A Method for Obtaining Digital Signatures and Public-Key Cryptosystems*. https://people.csail.mit.edu/rivest/Rsapaper.pdf

Sanger, David E. 2013. *Confront and Conceal: Obama's Secret Wars and Surprising Use of American Power*. New York: Broadway Books.

Singer, P.W., and Allan Friedman. 2013. *Cybersecurity and Cyberwar: What Everyone Needs to Know®*. New York: Oxford University Press.

Shearer, Jarrad. 2010. W32.stuxnet. https://www.symantec.com/security_response/writeup.jsp?docid=2010-071400-3123-99&tabid=2.

Shor, Peter W. 1991. Polynomial-Time Algorithms for Prime Factorization and Discrete Logarithms on a Quantum Computer. *SIAM Review* 41 (2): 303–332.

Spafford, H. 1989. Crisis and Aftermath. *Communications of the ACM* 32(6 June): 678–687. http://portal.acm.org/citation.cfm?id=63527.

Sparrow, R. 2007. Killer Robots. *Journal of Applied Philosophy* 24: 63–77.

Steinhoff, Uwe. 2013. Killing Them Safely: Extreme Asymmetry and Its Discontents. In *Killing by Remote Control: The Ethics of an Unmanned Military*, ed. B.J. Strawser. Oxford: Oxford University Press.

Turing, Alan M. 1939. *Turing's Treatise on the Enigma*. Unpublished Manuscript.

Chapter 8
Biological Sciences

Abstract Developments in the biological sciences have produced great benefits, including in relation to the control of diseases. However, in the recent and not so recent past, a number of governments have sought to develop biological weapons, e.g. the large-scale biological weapons program in the Soviet Union from 1946 to 1992. Moreover, there have been a number of acts, or attempted acts, of bioterrorism, notably by the Aum Shinrikyo in Japan. Techniques of genetic engineering have been available for some time to enhance the virulence, transmissibility and so on of naturally occurring pathogens. Recent developments in synthetic genomics have exacerbated the problem even further. Accordingly, there is the very real possibility of malevolent individuals or groups acquiring pathogens that have enhanced virulence and transmissibility and releasing them into the environment with catastrophic consequence. Some obvious regulatory measures that might be considered in relation to dual use issues include: regulations providing for mandatory physical safety and security of the storage, transport and physical access to samples of pathogens, equipment, laboratories etc.; mandatory licensing of dual-use technologies/techniques/pathogen samples; mandatory education and training; mandatory personnel security regulation e.g. background checks; censorship.

As mentioned in Chap. 2, Sect. 2.2, in the recent and not so recent past, a number of governments have sought to develop biological weapons, e.g. the large-scale biological weapons program in the Soviet Union from 1946 to 1992. Moreover, there have been a number of acts, or attempted acts, of bioterrorism, notably by the Aum Shinrikyo in Japan (they attempted to acquire and use anthrax and botulinum toxin), Al-Qaeda (they attempted to acquire and use anthrax) and the so-called Amerithrax attacks (involving the actual use of anthrax). Further, a small number of animal, plant and human pathogens are readily obtainable from nature, and bioterrorists with minimal microbiological training could use these to inflict causalities or economic damage.

Earlier versions of many of the claims, arguments and proposals in this chapter appeared in Miller and Selgelid (2007), Miller (2009), van der Bruggen et al. (2011) and Miller (2013).

© The Author(s) 2018
S. Miller, *Dual Use Science and Technology, Ethics and Weapons of Mass Destruction*, SpringerBriefs in Ethics, https://doi.org/10.1007/978-3-319-92606-3_8

Techniques of genetic engineering have been available for some time to enhance the virulence, transmissibility and so on of naturally occurring pathogens. Accordingly, there is the very real possibility of malevolent individuals or groups acquiring pathogens that have enhanced virulence and transmissibility and releasing them into the environment with catastrophic consequence. It should also be noted that the pathogens might be ones for which there are no vaccines, at least in the short term.

Recent developments in synthetic genomics have exacerbated the problem even further.[1] (See discussion in Chap. 2, Sect. 2.3 of ferret flu experiments.) It is now possible to create pathogens de novo, e.g. to produce deadly viruses from scratch. Accordingly, in the not too distant future it may be quite possible for a malevolent 'lone wolf' or member of a bioterrorist group with microbiological training to buy a bench-top DNA synthesizer and use it to assemble a specified genomic sequence of a highly virulent and transmissible pathogen from readily available raw materials.

There are, of course, impediments to such catastrophes. For instance, malevolent users would not have the ability to control the spread of epidemics that they instigate. So they would be putting their own lives and the lives of their supporters at risk. This inability would probably deter most malevolent individuals and groups, albeit perhaps not nihilistic 'end of the world' groups. Moreover, there is a lack of know-how and technological capacity concerning synthetic genomics amongst most researchers and laboratory workers. Finally, there are a plethora of safeguards in place, at least in the developed world, e.g., the US Select Agent regulations, albeit such 'lists' of pre-existing biological agents are not very effective against large numbers of newly created biological agents.[2]

In short, as we have seen in the case of chemical, nuclear and cyber R&D, *some* R&D in the biological sciences is dual use and, as such, has the potential for great harm, as well as great good. Moreover, there are a number of malevolent individuals and groups ready, willing and (increasingly) able to use this R&D to cause harm rather than to do good.

An important first step in responding the dual use dilemma in the biological sciences is to map the dual use terrain and the received method of mapping this terrain is by recourse to so-called experiments of concern. According to the National Research Council (NRC) report "experiments of concern" are those that would:

1. demonstrate how to render a vaccine ineffective;
2. confer resistance to therapeutically useful antibiotics or antiviral agents;
3. enhance the virulence of a pathogen or render a non-pathogen virulent;
4. increase the transmissibility of a pathogen;
5. alter the host range of a pathogen;
6. enable the evasion of diagnosis and/or detection by established methods; or
7. enable the weaponisation of a biological agent or toxin.[3]

---

[1] Selgelid (2016).

[2] Hence the need to turn to functional definitions of biological agents rather than lists of pre-existing ones.

[3] National Research Council (2004).

Other possible categories are:

8. Genetic sequencing of pathogens;
9. Synthesis of pathogenic micro-organisms;
10. Any experiment with *variola* virus (smallpox);
11. Attempts to recover/revive past pathogens.[4]

As noted in Chap. 2, and elsewhere in this book, the dual-use dilemma is a dilemma for multiple groups. It is obviously a dilemma for the researchers themselves. But it is also a dilemma for the institutions in which the researchers conduct their research and, in particular, those who manage these institutions. These institutions include universities and biotechnology companies. More generally, it is a dilemma for the individual communities for whose benefit or, indeed, to whose potential detriment, the research is being conducted, and for their national governments. After all, national governments have a moral responsibility in relation to the health and security of their citizens. Finally, given the global character of R&D in the biological sciences, not to mention of epidemics, the dual-use dilemma has become a dilemma for international bodies such as the United Nations.

In this chapter the emphasis is on an ethically informed regulatory response to dual use issues in the biological sciences and, specifically, on the collective responsibility to construct a web of prevention comprised of an integrated mix of regulatory measures. I begin with an account of the dual use issues arising in research based institutions, specifically, universities and commercial firms.

8.1 Research-Based Institutions

8.1.1 Universities and Scientific Freedom

As we saw in Chap. 4, Sect. 4.1, the scientific enterprise is a cooperative endeavour that is an intrinsic good (collective expert knowledge being an end-in-itself) and one that is best conducted under conditions of scientific freedom and scientific openness; or, at least, this is the received view. Accordingly, governmental control of research and censorship are anathema to the scientific enterprise. Nowhere is this conception of science more strongly held than in universities. In the universities scientific freedom, and the cognate value of academic freedom, are core institutional values.

It is worth spelling out some of ethical issues here. In what follows I provide a brief analysis of academic freedom. The argument for the principle of academic freedom begins with the premise that freedom of intellectual inquiry is a fundamental human right.[5]

Thus conceived, freedom of intellectual inquiry is not an individual right of the ordinary kind. Although it is a right which attaches to individuals, as opposed to

---

[4]Miller and Selgelid (2008).

[5]An earlier version of much of the material in this section appeared in Miller (2000, 110–131).

groups, it is not a right which an individual could exercise by him/herself. Communication, discussion and inter-subjective methods of testing are social, or at least interpersonal, activities. However, it is important to stress that they are not activities which are relativised to social or ethnic or political groups; in principle, intellectual interaction can and ought to be allowed to take place between individuals irrespective of whether they belong to the same social, ethnic or political group. In short, freedom of intellectual inquiry, or at least its constituent elements, is a fundamental *human* right. Note that being a fundamental human right it can, at least in principle, sometimes override collective interests and goals including organisational, and even national, economic interests and goals. This 'trumping' property of human rights is a constitutive element of liberal democracy; a form of polity whose legitimacy is based in part on its capacity and willingness to protect human rights including, at times, against infringements emanating from the government of the day.

If freedom of intellectual inquiry is a human right then like other human rights, such as the right to life and to freedom of the person, it is a right which academics as humans possess along with all other citizens. But how does this bear upon the specific institutional purpose of the university to acquire, transmit and disseminate knowledge? Before we can answer this question we need to get clearer on the relationship between the human right to freely engage in intellectual inquiry on the one hand, and knowledge or truth on the other.

Freedom of intellectual inquiry and knowledge are not simply related as means to end, but also conceptually. To freely inquire is to seek the truth by reasoning. Truth is not an external contingently connected end which some inquiries might be directed towards if the inquirer happened to have an interest in truth, rather than, say, an interest in falsity. Rather truth is internally connected to intellectual inquiry. An intellectual inquiry which did not aim at the truth would not be an intellectual inquiry, or at least would be defective qua intellectual inquiry. Moreover, here aiming at truth is aiming at truth as an end in itself. (This is not inconsistent with also aiming at truth as a means to some other end.) Further, to engage in free intellectual inquiry in our extended sense involving communication with, and testing by, others, is to freely seek the truth by reasoning with others. Intellectual inquiry in this sense is not exclusively the activity of a solitary individual.

Given that freedom of intellectual inquiry is a human right, and given the above described relationship between intellectual inquiry and truth (or knowledge) we can now present the argument in relation to freedom of intellectual inquiry. This argument in effect seeks to recast the notion of freedom of intellectual inquiry in order to bring out the potential significance for conceptions of the university of the claim that freedom of intellectual inquiry is a human right.

1. Freedom of intellectual inquiry is a human right.
2. Freedom of intellectual inquiry is (principally) freedom to seek the truth by reasoning with others.
3. Freedom to seek the truth by reasoning with others is a fundamental human right.

Let us grant the existence of a human right to freely pursue the truth by reasoning with others. What are the implications of this right for universities and for academics' freedom of inquiry? Given such a right of intellectual inquiry, it is plausible to conclude that the university is simply the institutional embodiment of that moral right. In short, the university is the institutional embodiment of the right to freely seek the truth by reasoning with others. The following claims now seem warranted.

First, universities have been established as centres wherein independence of intellectual inquiry is maintained. This flows from the proposition that the university is an institutional embodiment of the moral *right* of the inquirers to freely undertake their intellectual inquiries. Universities are not, for example, research centres set up to pursue quite specific intellectual inquiries determined by their external funders. Nor should particular inquiries undertaken by academics at universities be terminated on the grounds that some external powerful group, say government, might not find the truths discovered in the course of these inquiries politically palatable.

Second, universities have a duty to disseminate scholarship and research to the community. Intellectual inquiry is not only a human right, it is an activity which produces external benefits. For example, knowledge is a means to other goods, including economic well-being. Accordingly, and notwithstanding the rights of academics to freely inquire, it is reasonable that, qua community supported institutions, universities take on an obligation to ensure that their intellectual activities have a flow through effect to the wider community in terms of such external benefits. Thus dissemination of research (usually) has obvious benefits to the community, including health and economic benefits.

On the view of the university under consideration, interference in the process of the free pursuit of knowledge in universities strikes at one of the fundamental purposes for which universities have been established. Such interference could not be justified, for example, on the grounds that whereas free inquiry might be necessary for the acquisition of knowledge in many instances, in some particular instance free inquiry was not leading to knowledge, and therefore in this case free inquiry could be interfered with without striking at the basic purposes of the university as an institution.

Moreover, the university, in so far as it pursues this purpose, can so pursue it, even if so doing is inconsistent with the collective goals and interests of the community or government. In this respect the right of intellectuals to pursue the truth is akin to the right of the judiciary to pursue justice even in the face of conflicting collective goals and interests, including the national interest. For example, European academics researching political or ethical issues in say, China, have a right to publish that research notwithstanding the damage it might do to present diplomatic relations and economic prospects.

Notwithstanding the importance of the human right of intellectual inquiry and its centrality to the institution of the university, freedom of intellectual inquiry in general, and of scientific inquiry in particular is not an absolute right. Specifically, it can be overridden if its exercise comes into conflict with other human rights, notably the right to life. Accordingly, if a contingency arose, such as war or a pandemic or a potential terrorist attack, then the duty of a scientist to disseminate her findings could well be overridden. Doubtless, in relation to most academic research such contingencies are exceptions, and should be treated as such. Nevertheless, given the high risk to human life and health posed by misuse of research in synthetic biology and related areas, such biological research constitutes a special case. Censorship of academic research needs special justification. However, that justification is, in general terms, available in the areas in question, e.g. the high risk of misuse by an extremist 'end of the world' seeking group of such research. Naturally, censorship of any specific research or research project will not only need some justification, it will need a *specific* justification that details the high risk of misuse of this specific research project outcome by some malevolent group or individual, e.g. the research outcome is a highly virulent, easily transmissible and readily weaponised pathogen and, therefore, the potential harm is disproportionately great relative to the projected benefits.

In addition to the argument for free and open scientific research based on the right to intellectual, academic and/or scientific freedom there is the argument from scientific progress; academic freedom, including freedom of dissemination, is necessary for scientific progress. While this argument has a certain amount of weight it is far from decisive, given that scientific freedom and openness is a matter of degree and, relatedly, given that the necessity in question is not strict necessity. (Here, as elsewhere, the notion of necessity is the relatively weak notion of 'necessary' means to an end rather than the much stronger notion of logically necessity.) Let me explain. Scientific freedom and openness can be curtailed to some extent without halting the progress of science. Indeed, scientific progress in nuclear physics appears to be a case in point, given the long history of secrecy and censorship in the nuclear industry. Moreover, trade secrets in the chemical industry and, for that matter, biotechnology appear to be consistent with scientific progress.

One might admit that science has suffered from these admittedly real constraints on freedom but nonetheless claim that science would have advanced *even further than it has* if there had been more freedom and openness in science than has actually been the case. Even if correct, however, this would not go to show that no restrictions on the dissemination of scientific information are warranted. The progress of science is just one of many legitimate social aims that must be taken into consideration by scientists and policy makers alike. The progress of science is important—as is the human right to freedom of inquiry and the institutional right to academic freedom—but other things such as public health/security are important too; and there is no compelling reason to think that these two kinds of goals will never conflict or that the former should always be given absolute priority over the latter (or vice-versa), in cases of conflict, regardless of the extent to which the latter is threatened.

The common-sense view mentioned above is that if trade-offs need to be made between, say, rights to disseminate and scientific progress on the one hand, and security/public health needs on the other, then a reasonable balance should be struck between these on the basis of the varying *moral weight* to be accorded to each—and taking into account the *probability* of the harmful versus beneficial outcomes of the competing courses of action.

8.1.2 Commercial Firms

Private sector research raises particular issues in relation to the dual-use dilemma.[6] The biotechnology industry is important both economically and in terms of the development of the genetic sciences. However, industrial research is primarily and often explicitly motivated by the pursuit of profits rather than, for instance, by knowledge for its own sake. Naturally, as is the case with university-based research, there are mixed motives. Nevertheless, it can be assumed that the profit motive is considerably stronger in commercial firms than in universities, even if the contrast should not be overdrawn. This has a number of implications.

Firstly, industrial discoveries are often kept secret rather than published or otherwise widely shared with the academic community. Accordingly, there is less public awareness and scrutiny of dual use research. Second, as noted in Chap. 4, the profit motive might unduly influence decisions regarding dual use research; decision might be overly permissive. Thirdly, private sector research is not generally subject to the same degree of institutional oversight (via institutional ethics committees) as that which takes place in universities.

In summation, while the private sector is very important in terms of R&D in the biological sciences it has not adequately addressed dual use issues. Indeed, according to Corneliussen:

"Biotechnology is exploited most intensively in commercial enterprises. This is also where the potential for misuse is most acute, as a result of heavy investments in both intellectual property and highly specialized equipment. In terms of industry, [a Sunshine Project study conducted in 2004] found that only about 70 firms had NIH registered institutional biosafety/biosecurity committees (IBCs). According to Estimates by Ernst and Young (2005), the US biotechnology industry comprises about 1500 companies. Not all of these conduct recombinant DNA research; nevertheless, 70 seems an unexpectedly low figure. Of the 70 firms, only 26 responded to the [Sunshine Project] survey, 14 of which provided minutes. None of the minutes were deemed to be adequate…. The survey further revealed that some private sector IBCs did not review specific research projects, but instead issued blanket approvals without regard for individual project details…. The same report cites Merck as stating "We currently do not perform any research or manufacturing that requires IBC review, the committee has therefore been dissolved" [and cites other companies such as Hoffman-La Roche and IDEC saying similar things]."[7]

8.2 Regulation

In light of the discussion in Sect. 8.1, it is clear that there is a need for the implementation of a significant array of regulatory measures in the biological sciences in

---

[6]An earlier version of the material in this section appeared in Miller and Selgelid (2011, 1–122).

[7]Corneliussen (2006, S50–S52).

respect of dual use issues, albeit not over-regulation to the point that, so to speak, the baby is thrown out with the bathwater. Moreover, those with responsibilities in this regard are many. They include not only researchers, but institutional managers, members of governments and, for that matter, citizens. Further, some of these measures may include ensuring collective public ignorance or, at least, constraining collective expert knowledge in relation to various technical issues in the biological sciences. Accordingly, and given what is at stake, there is a collective moral responsibility in this regard. As we have seen in earlier chapters, these measures ought to be integrated such that taken together they constitute a web of prevention; so the collective moral responsibility is to be institutionally embedded in a web of prevention comprised of various regulatory measures. Such measures include the imposition of limits on dual-use experiments and on the dissemination of potentially dangerous information resulting from dual-use discoveries. These measures themselves exist on a spectrum ranging from the least intrusive/restrictive to the most intrusive/restrictive.

Some obvious regulatory measures that might be considered include the following ones.

(1) Mandatory Physical Safety and Security Regulation: regulations providing for mandatory physical safety and security of the storage, transport and physical access to samples of pathogens, equipment, laboratories etc.

(2) Licensing of Dual-Use Technologies/Techniques: mandatory licensing of dual-use technologies/techniques/pathogen samples. Only certain laboratories in the public sector and the private sector might be licensed to engage in research involving the use of certain dual-use technologies and licenses for DNA synthesizers might be required.

(3) Mandatory Education and Training: Given the potential harms arising from, for example, the identified types of experiments of concern it is clear that some process of education and/or training for relevant researchers and other personnel is called for.

(4) Mandatory Personnel Security Regulation: Doubtless it is prudent, indeed it is a moral requirement, that access to virulent pathogens be disallowed to a researcher diagnosed as a psychopath or to a known member of a terrorist organisation.

(5) Censorship/Constraint of Dissemination: The question of whether research findings ought to be freely disseminated, censored or their dissemination in some lesser way restricted is an extremely difficult issue and it is by no means obvious who the ultimate decision-maker ought to be. A relevant important distinction here is that between 1st tier and 2nd tier dual-use research. For example, 1st tier research findings might need to be disseminated in such a way that anyone being informed of these findings would not be able to replicate the experiments that enabled the results reported in the findings.

Let us assume that a range of regulatory options ought to be pursued, both at the institutional (university, commercial firm, government research laboratory), national and international levels (e.g. Biological Weapons Convention verification processes). Moreover, these options need to embrace ones tailored to security and safety concerns beyond dual use issues. While the focus of this book is on dual use issues, these exist on a spectrum of security and safety concerns and the regulatory architecture needs to thought of in holistic terms; it does not make sense to develop a regulatory architecture exclusively for all dual use issues, given the latter straddle both safety and security concerns and, more generally, safety and security issues infect one another. There remain some in principle obstacles to the establishment of adequate measures to deal with the dual use problem in science and technology and a number of these stem from various perverse incentive structures that derive from collective action problems identified in Sect. 8.1.1 and in Chap. 4, Sect. 4.4. For reasons of space I cannot here pursue these matters further. Rather I conclude this section and, indeed, this chapter, by outlining a set of specific recommendations that have been put forward in order to deal with dual use issues in the biological sciences[8]:

(1) Mandatory awareness raising, training, and education, as required.
(2) Extend the remit of existing biosafety committees in universities to include biosecurity issues (in part in light of the fact that dual use issues have a biosafety as well as a biosecurity dimension by virtue of the potential for culpable negligence [Chap. 2 Sect. 2.1]).
(3) Require commercial firms to establish biosafety/biosecurity committees.
(4) Develop enforceable professional codes of conduct for relevant personnel (e.g., scientists).
(5) At the national level, establish an independent authority to deal with safety and security issues in biological/converging sciences including, but not restricted to, dual use issues.
(6) Select agent rules should be revised to be based on functionality rather than lists of agents.
(7) Governments should make/implement an international/multilateral agreement regarding safety and security issues in the biological sciences, including but not restricted to, dual use issues.
(8) National legislation and protocols regarding safety and security issues in the biological sciences, including but not restricted to dual use issues, should be standardised and harmonised. (E.g., export controls should apply worldwide.)
(9) Control over buying and selling of DNA sequences and/or other dual use materials should be overseen by an international clearinghouse established under a multilateral agreement (i.e., all orders would need to be reported to, and approved by, the clearinghouse).
(10) Verification procedures should be added to the Biological and Toxin Weapons Convention.
(11) Dual use measures will be periodically reviewed/revised at the institutional, national, and international level.

---

[8]Miller and Selgelid (2011, 84–85).

8.3   Conclusion

In this chapter I have discussed dual use issues in the biological sciences and elaborated on the tension between the importance of scientific freedom and the benefits of R&D in biotechnology, on the one hand, and the potential for harm (e.g. if malevolent groups or individuals have access to research that has led to the creation of highly virulent pathogens transmissible to humans), on the other. Importantly, I have argued the case for a so-called web of prevention and identified some of its key regulatory components.

References

Corneliussen, Filippa. 2006. Adequate Regulation, a Stop-Gap Measure, or Part of a Package? *EMBO Reports* 7: S50–S52.

Miller, Seumas. 2000. Academic Autonomy. In *Why Universities Matter*, ed. C.A.J. Coady, 110–131. Sydney: Allen and Unwin.

Miller, Seumas. 2009. *Terrorism and Counter-Terrorism: Ethics and Liberal Democracy*. Oxford: Blackwell.

Miller, Seumas. 2013. Moral Responsibility, Collective Action Problems and the Dual Use Dilemma in Science and Technology. In *On the Dual Uses of Science and Ethics*, ed. Michael Rappert, and Brian Selgelid. Canberra: ANU Press.

Miller, Seumas, and Michael Selgelid. 2007. Ethical and Philosophical Consideration of the Dual Use Dilemma in the Biological Sciences. *Science and Engineering Ethics* 13: 523–580.

Miller, Seumas, and Michael Selgelid. 2008. *Ethical and Philosophical Consideration of the Dual-Use Dilemma in the Biological Sciences*. Dordrecht: Springer.

Miller, Seumas and Michael Selgelid. 2011. *Report on Biosecurity and Dual Use Research (Part 2)*, 1–122 and 84–85. The Hague: Dutch Research Council.

Selgelid, Michael. 2016. Gain of Function Research: Ethical Analysis. *Science and Engineering Ethics* 22 (4): 923–964.

National Research Council. 2004. *Biotechnology Research in an Age of Terrorism*. Washington, DC: National Academies Press.

van der Bruggen, Koos, Seumas Miller and Michael Selgelid. 2011. *Report on Biosecurity and Dual Use Research*, 1–122. Dutch Research Council.

Chapter 9
Conclusion

Abstract The main arguments and findings in this work are summarised. Dual use issues are to be found in the chemical industry, nuclear industry, in cyber-technology and in the biological sciences. Moreover, they are exacerbated by collective action problems. However, they exist in a somewhat different form in different domains of science and technology, (e.g. nuclear vs. biological sciences), and in somewhat diverse institutional settings (e.g. universities vs. private firms). Therefore, the appropriate responses to the problem in these different domains of science and technology and different institutional settings may need to differ somewhat. That said, these domains and institutions do share some common general features. Firstly, in each case the dual use issues in question may call for restrictions on R&D research and dissemination of findings; something that is, generally speaking, antithetical to scientists and technologists. Secondly, they are a collective moral responsibility, e.g. of scientists and governments. Thirdly, the response needs to be multi-faceted and will typically involve a so-called 'web of prevention' (an integrated suite of regulatory measures).

The problem of dual-use science research and technology arises because such research and technology has the potential to be used for great evil as well as for great good. In this work dual use issues have been considered in the chemical industry, nuclear industry, cyber-technology and the biological sciences. On my somewhat stipulative definitional account new and emerging science or technology is dual use if:

(1) It can be used for both beneficial and harmful purposes—where either the harmful purposes involve the use of weapons as means, and usually weapons of mass destruction in particular, or serious, large-scale harm that does not necessarily involve weaponisation;

(2) The serious, large-scale harm in question is caused by a single act of using the technology—as opposed to multiple acts that in aggregate cause great harm;

(3) A large-scale beneficial outcome is intended by the original researchers;

© The Author(s) 2018

S. Miller, *Dual Use Science and Technology, Ethics and Weapons of Mass Destruction*, SpringerBriefs in Ethics, https://doi.org/10.1007/978-3-319-92606-3_9

(4) The actual or potential harmful outcome is reasonably foreseeable by the original researchers and, if it eventuates, is either intended by secondary malevolent users or, at least, their secondary use involves culpable negligence.

I have distinguished and analysed various categories of collective knowledge and collective ignorance and argued that scientists and technologists—jointly with members of governments etc.—have various collective moral responsibilities in this regard. For instance—and somewhat paradoxically—they have a collective moral responsibility to maintain or bring about collective *public ignorance* with respect to certain aspects of dual use knowledge and certainly with respect to how to make WMDs. I have also argued that there is a collective moral responsibility to try to ensure that no person (or as few persons as possible)—whether expert or not—individually knows how to make a WMD and (obviously) to try to ensure that malevolent groups do not have the collective expert knowledge to make WMDs. From this certain things follow, or so I have argued. For instance, I have argued that scientists and technologists must accept a collective responsibility (jointly with legislators etc.) to design and implement training programs, regulations and so on to deal with dual use issues.

I have defined the notion of collective moral responsibility as follows: if agents, A, B, C etc. are naturally or institutionally responsible for a joint (including epistemic) activity x (and/or some foreseeable outcome of x, O) and x (and/or O) is morally significant then—other things being equal—A, B, C etc. are collectively (i.e. jointly) morally responsible for x (and/or O) and—other things being equal—can be praised or blamed for x (and/or O).

In relation to the chemical industry and the biological sciences in particular, it has been argued that the responsibility to engage in dual use harm prevention in respect of R&D is a collective moral responsibility; specifically, a collective moral responsibility to design and implement an institutionally-based *web of prevention*. This web of prevention consists of the CWC and a strengthened BWC (respectively) and a range of additional regulations, governance arrangements and the like. These include restrictions on export of toxins and pathogens, and prescribing of the safety and security-based conditions under which R&D can be undertaken and by whom, e.g. background checks and security clearance for research personnel, training programs, licensing of organisations, and so on.

My definition of dual use technology was applied to cyber-technology. It turns out that computer viruses may well be instances of dual use technology. Importantly, according to this definition, cyber-technology used to effect mass destruction and in which the weapons used are controlled by computers, including with respect to the selection of targets (and, perhaps the selection of the weapons themselves), i.e. autonomous weaponry, also constitutes dual use technology. It is also suggested that various forms of ransom-ware constitute dual use technology.

The dual use problem exists in an acute form in the nuclear industry. For scientific research, technology and materials in the nuclear sciences have enabled, on the one hand, unbounded nuclear energy for peaceful purposes, and yet on the other, massive arsenals of nuclear WMDs with the potential to destroy humankind. Moreover, the nuclear arms race is a collective action problem. The generic solution to this kind of collective action problem in the nuclear sciences is at least in part an enforced cooperative scheme: enforced cooperative schemes are one important way to embed collective moral responsibility in institutional settings suffering from harm inducing collective action problems. This requires widening and strengthening existing institutional arrangements such as the NPT, but also creating additional ones, especially in the area of enforcement.

Index

© The Author(s) 2018
S. Miller, *Dual Use Science and Technology, Ethics and Weapons of Mass
Destruction*, SpringerBriefs in Ethics, https://doi.org/10.1007/978-3-319-92606-3

.

The manufacturer's authorised representative in the EU is Springer
Nature Customer Service Centre GmbH, Europaplatz 3, 69115 Heidelberg,
Germany. If you have any concerns regarding our products, please
contact ProductSafety@springernature.com

Printed and bound by CPI Group (UK) Ltd, Croydon, CR0 4YY
27/04/2026
02097632-0007